NIC 2004-13

# Mapping the Global Future

## Report of the National Intelligence Council's 2020 Project

*Based on consultations with nongovernmental experts around the world*

*December 2004*

**To obtain a copy of this publication, please contact:**
Government Printing Office (GPO), Superintendent of Documents, PO Box 391954,
Pittsburgh, PA 15250-7954; Phone: (202) 512-1800; Fax: (202) 512-2250; http:\\bookstore.gpo.gov;
GPO Stock 041-015-0024-6; ISBN 0-16-073218-2.

# NATIONAL INTELLIGENCE COUNCIL
## Washington, D.C. 20505

**From the Chairman of the National Intelligence Council**

*Mapping the Global Future:  Report of the National Intelligence Council's 2020 Project* is the third unclassified report prepared by the National Intelligence Council (NIC) in recent years that takes a long-term view of the future.  It offers a fresh look at how key global trends might develop over the next decade and a half to influence world events. Mindful that there are many possible "futures," our report offers a range of possibilities and potential discontinuities, as a way of opening our minds to developments we might otherwise miss.

As I used to say to my students at Princeton, linear analysis will get you a much-changed caterpillar, but it won't get you a butterfly. For that you need a leap of imagination. We hope this project, and the dialogue it stimulates, will help us make that leap—not to *predict* the world of 2020, which is clearly beyond our capacity—but to better *prepare* for the kinds of challenges that may lie ahead.

*Mapping the Global Future* builds upon methods used to develop our two earlier studies by employing a variety of innovative methodologies and approaches, including extensive consultations with a wide range of governmental and nongovernmental experts.

- The *Global Trends 2010* paper was derived from a series of conferences held in the Washington, DC area, attended by academic and business leaders who conferred with Intelligence Community experts.  Produced in 1997, it was the centerpiece of numerous briefings to policymakers.

- *Global Trends 2015*, an ambitious and ground-breaking effort, identified seven key drivers of global change:  demographics, natural resources and the environment, science and technology, the global economy and globalization, national and international governance, future conflict, and the role of the United States.  Produced in December 2000, it was based upon discussions between the National Intelligence Council and a broad array of nongovernmental specialists in the United States.  *GT 2015* received international attention and prompted a lively debate about the forces that will shape our world. We billed it as "a work in progress, a flexible framework for thinking about the future that we will update and revise as conditions evolve."

*Mapping the Global Future* picks up where *Global Trends 2015* left off but differs from our earlier efforts in three principal respects:

- *We have consulted experts from around the world in a series of regional conferences to offer a truly global perspective.*  We organized conferences on five continents to solicit the views of foreign experts on the prospects for their regions over the next 15 years.

- *We have relied more on scenarios to try to capture how key trends might play out.*  Our earlier efforts focused on key trends that would impact regions and key countries of interest.

# NATIONAL INTELLIGENCE COUNCIL

The trends we highlight in this paper provide a point of departure for developing imaginative global scenarios that represent several plausible alternative futures.

- ***We have developed an interactive Web site to facilitate an ongoing, global dialogue.*** The Web site also contains links to a wealth of data of interest to scholars and the general public.

The entire process, from start to finish, lasted about a year and involved more than a thousand people. We appreciate the time and effort that each contributed to ***Mapping the Global Future.*** The Methodology section of this report acknowledges the special contributions of individual scholars and organizations and the many conferences and symposia held in conjunction with the project. Within the NIC, Craig Gralley, Director of Strategic Plans and Outreach, deserves special mention for his management of the many dozens of conferences, workshops, and planning sessions associated with the project. Let me also extend special recognition to Mathew Burrows, Director of the NIC's Analysis and Production Staff, who with creativity and clarity brought together the disparate parts of the project into an elegant final draft. Elizabeth Arens and Russell Sniady, members of his staff, also made significant contributions.

I encourage readers to review the complete set of 2020 Project documents found on the National Intelligence Council's Web site, www.cia.gov/nic, and to explore the scenario simulations. We continue to see this project as a work in progress—a way of catalyzing an ongoing dialogue about the future at a time of great flux in world affairs.

Robert L. Hutchings

# Contents

|  | Page |
|---|---|
| **Executive Summary** | 9 |
| **Methodology** | 19 |
| **Introduction** | 25 |
| **The Contradictions of Globalization** | 27 |
| An Expanding and Integrating Global Economy | 29 |
| The Technology Revolution | 34 |
| Lingering Social Inequalities | 37 |
| *Fictional Scenario: Davos World* | 40 |
| **Rising Powers: The Changing Geopolitical Landscape** | 47 |
| Rising Asia | 48 |
| Other Rising States? | 51 |
| The "Aging" Powers | 56 |
| Growing Demands for Energy | 59 |
| US Unipolarity—How Long Can It Last? | 63 |
| *Fictional Scenario: Pax Americana* | 64 |
| **New Challenges to Governance** | 73 |
| Halting Progress on Democratization | 73 |
| Identity Politics | 79 |
| *Fictional Scenario: A New Caliphate* | 83 |
| **Pervasive Insecurity** | 93 |
| Transmuting International Terrorism | 93 |
| Intensifying Internal Conflicts | 97 |
| Rising Powers: Tinder for Conflict? | 98 |
| The WMD Factor | 100 |
| *Fictional Scenario: Cycle of Fear* | 104 |
| **Policy Implications** | 111 |

## Graphics and Tables

China's and India's Per Capita GDPs Rising Against US    31

When China's and India's GDPs Would Exceed Today's Rich Countries    32

Telescoping the Population of the World to 2020    48

China's Rise    50

Projected Rise in Defense Spending, 2003–2025    51

Fossil Fuels Will Continue to Dominate in 2020    59

An Expanding European Union    60

Number of Religious Adherents, 1900–2025    80

Key Areas of Radical Islamic Activities Since 1992    82

EU: Estimated and Projected Ratios of Muslims to Ethnic Europeans, 1985-2025    83

Global Trends in Internal Conflict, 1990-2003    101

## Special Topics

The 2020 Global Landscape   8

Mapping the Global Future   26

What Would an Asian Face on Globalization Look Like?   28

What Could Derail Globalization?   30

Biotechnology: Panacea and Weapon   36

The Status of Women in 2020   38

Risks to Chinese Economic Growth   52

India vs. China: Long-Term Prospects   53

Asia: The Cockpit for Global Change?   55

Global Aging and Migration   58

Could Europe Become A Superpower?   61

The Geopolitics of Gas   62

Eurasian Countries: Going Their Separate Ways?   74

Climate Change and Its Implications Through 2020   76

Latin America in 2020: Will Globalization Cause the Region to Split?   78

Organized Crime   96

Cyber Warfare?   97

How Can Sub-Saharan Africa Move Forward?   99

International Institutions in Crisis   102

The Rules of War: Entering "No Man's Land"   103

Post-Combat Environments Pose the Biggest Challenge   104

Is the United States' Technological Prowess at Risk?   112

How the World Sees the United States   114

# The 2020 Global Landscape

| Relative Certainties | Key Uncertainties |
|---|---|
| Globalization largely irreversible, likely to become less Westernized. | Whether globalization will pull in lagging economies; degree to which Asian countries set new "rules of the game." |
| World economy substantially larger. | Extent of gaps between "haves" and "have-nots"; backsliding by fragile democracies; managing or containing financial crises. |
| Increasing number of global firms facilitate spread of new technologies. | Extent to which connectivity challenges governments. |
| Rise of Asia and advent of possible new economic middle-weights. | Whether rise of China/India occurs smoothly. |
| Aging populations in established powers. | Ability of EU and Japan to adapt work forces, welfare systems, and integrate migrant populations; whether EU becomes a superpower. |
| Energy supplies "in the ground" sufficient to meet global demand. | Political instability in producer countries; supply disruptions. |
| Growing power of nonstate actors. | Willingness and ability of states and international institutions to accommodate these actors. |
| Political Islam remains a potent force. | Impact of religiosity on unity of states and potential for conflict; growth of jihadist ideology. |
| Improved WMD capabilities of some states. | More or fewer nuclear powers; ability of terrorists to acquire biological, chemical, radiological, or nuclear weapons. |
| Arc of instability spanning Middle East, Asia, Africa. | Precipitating events leading to overthrow of regimes. |
| Great power conflict escalating into total war unlikely. | Ability to manage flashpoints and competition for resources. |
| Environmental and ethical issues even more to the fore. | Extent to which new technologies create or resolve ethical dilemmas. |
| US will remain single most powerful actor economically, technologically, militarily. | Whether other countries will more openly challenge Washington; whether US loses S&T edge. |

## Executive Summary

**At no time since the formation of the Western alliance system in 1949 have the shape and nature of international alignments been in such a state of flux.** The end of the Cold War shifted the tectonic plates, but the repercussions from these momentous events are still unfolding. Emerging powers in Asia, retrenchment in Eurasia, a roiling Middle East, and transatlantic divisions are among the issues that have only come to a head in recent years. The very magnitude and speed of change resulting from a globalizing world—apart from its precise character—will be a defining feature of the world out to 2020. Other significant characteristics include: the rise of new powers, new challenges to governance, and a more pervasive sense of insecurity, including terrorism. As we map the future, the prospects for increasing global prosperity and the limited likelihood of great power conflict provide an overall favorable environment for coping with what are otherwise daunting challenges. **The role of the United States will be an important variable in how the world is shaped, influencing the path that states and nonstate actors choose to follow.**

## New Global Players

**The likely emergence of China and India, as well as others, as new major global players—similar to the advent of a united Germany in the 19th century and a powerful United States in the early 20th century—will transform the geopolitical landscape, with impacts potentially as dramatic as those in the previous two centuries.** In the same way that commentators refer to the 1900s as the "American Century," the 21st century may be seen as the time when Asia, led by China and India, comes into its own. A combination of sustained high economic growth, expanding military capabilities, and large populations will be at the root of the expected rapid rise in economic and political power for both countries.

- Most forecasts indicate that by 2020 China's gross national product (GNP) will exceed that of individual Western economic powers except for the United States. India's GNP will have overtaken or be on the threshold of overtaking European economies.

- Because of the sheer size of China's and India's populations—projected by the US Census Bureau to be 1.4 billion and almost 1.3 billion respectively by 2020—their standard of living need not approach Western levels for these countries to become important economic powers.

Barring an abrupt reversal of the process of globalization or any major upheavals in these countries, the rise of these new powers is a virtual certainty. Yet how China and India exercise their growing power and whether they relate cooperatively or competitively to other powers in the international system are key uncertainties. The economies of other developing countries, such as Brazil, could surpass all but the largest European countries by 2020; Indonesia's economy could also approach the economies of individual European countries by 2020.

By most measures—market size, single currency, highly skilled work force, stable democratic governments, and unified trade bloc—an enlarged Europe will be able to increase its weight on the international scene. Europe's strength could be in providing a model of global and regional governance to the rising powers. But aging populations and shrinking work forces in most countries will have an important impact on the continent. Either European countries adapt their work forces, reform their social welfare, education, and tax systems, and accommodate growing immigrant populations (chiefly from Muslim countries), or they face a period of protracted economic stasis.

Japan faces a similar aging crisis that could crimp its longer run economic recovery, but it also will be challenged to evaluate its regional status and role. Tokyo may have to choose between "balancing" against or "bandwagoning" with China. Meanwhile, the crisis over North Korea is likely to come to a head sometime over the next 15 years. Asians' lingering resentments and concerns over Korean unification and cross-Taiwan Strait tensions point to a complicated process for achieving regional equilibrium.

Russia has the potential to enhance its international role with others due to its position as a major oil and gas exporter. However, Russia faces a severe demographic crisis resulting from low birth rates, poor medical care, and a potentially explosive AIDS situation. To the south, it borders an unstable region in the Caucasus and Central Asia, the effects of which—Muslim extremism, terrorism, and endemic conflict—are likely to continue spilling over into Russia. While these social and political factors limit the extent to which Russia can be a major global player, Moscow is likely to be an important partner both for the established powers, the United States and Europe, and for the rising powers of China and India.

With these and other new global actors, **how we mentally map the world in 2020 will change radically**. The "arriviste" powers—China, India, and perhaps others such as Brazil and Indonesia—have the potential to render obsolete the old categories of East and West, North and South, aligned and nonaligned, developed and developing. Traditional geographic groupings will increasingly lose salience in international relations. A state-bound world and a world of mega-cities, linked by flows of telecommunications, trade and finance, will co-exist. Competition for allegiances will be more open, less fixed than in the past.

## Impact of Globalization

**We see globalization**—growing interconnectedness reflected in the expanded flows of information, technology, capital, goods, services, and people throughout the world—**as an overarching "mega-trend," a force so ubiquitous that it will substantially shape all the other major trends in the world of 2020**. But the future of globalization is not fixed; states and nonstate actors—including both private companies and NGOs—will struggle to shape its contours. Some aspects of globalization—such as the growing global interconnectedness stemming from the information technology (IT) revolution— almost certainly will be irreversible. Yet it is also possible, although unlikely, that the process of globalization could be slowed or even stopped, just as the era of globalization in the late 19<sup>th</sup> and early 20<sup>th</sup> centuries was reversed by catastrophic war and global depression.

Barring such a turn of events, **the world economy is likely to continue growing impressively:  by 2020, it is projected to be about 80 percent larger than it was in 2000, and average per capita income will be roughly 50 percent higher.** Of course, there will be cyclical ups and downs and periodic financial or other crises, but this basic growth trajectory has powerful momentum behind it.  Most countries around the world, both developed and developing, will benefit from gains in the world economy.  By having the fastest-growing consumer markets, more firms becoming world-class multinationals, and greater S&T stature, Asia looks set to displace Western countries as the focus for international economic dynamism—provided Asia's rapid economic growth continues.

**Yet the benefits of globalization won't be global.** Rising powers will see exploiting the opportunities afforded by the emerging global marketplace as the best way to assert their great power status on the world stage.  In contrast, some now in the "First World" may see the closing gap with China, India, and others as evidence of a relative decline, even though the older powers are likely to remain global leaders out to 2020.  The United States, too, will see its relative power position eroded, though it will remain in 2020 the most important single country across all the dimensions of power.  Those left behind in the developing world may resent China and India's rise, especially if they feel squeezed by their growing dominance in key sectors of the global marketplace.  And large pockets of poverty will persist even in "winner" countries.

**The greatest benefits of globalization will accrue to countries and groups that can access and adopt new technologies.** Indeed, a nation's level of technological achievement generally will be defined in terms of its investment in *integrating and applying* the new, globally available technologies—whether the technologies are acquired through a country's own basic research or from technology leaders.  The growing two-way flow of high-tech brain power between the developing world and the West, the increasing size of the information computer-literate work force in some developing countries, and efforts by global corporations to diversify their high-tech operations will foster the spread of new technologies.  High-tech breakthroughs—such as in genetically modified organisms and increased food production—could provide a safety net eliminating the threat of starvation and ameliorating basic quality of life issues for poor countries.  But the gap between the "haves" and "have-nots" will widen unless the "have-not" countries pursue policies that support application of new technologies— such as good governance, universal education, and market reforms.

Those countries that pursue such policies could leapfrog stages of development, skipping over phases that other high-tech leaders such as the United States and Europe had to traverse in order to advance.  **China and India are well positioned to become technology leaders, and even the poorest countries will be able to leverage prolific, cheap technologies to fuel—although at a slower rate—their own development.**

- The expected next revolution in high technology involving the convergence of nano-, bio-, information and materials technology could further bolster China and India's prospects. Both countries are investing in basic research in these fields and are well placed to be leaders in a number of key fields. Europe risks slipping behind Asia in some of these technologies. The United States is still in a position to retain its overall lead, although it must increasingly compete with Asia to retain its edge and may lose significant ground in some sectors.

**More firms will become global, and those operating in the global arena will be more diverse, both in size and origin, more Asian and less Western in orientation. Such corporations, encompassing the current, large multinationals, will be increasingly outside the control of any one state and will be key agents of change in dispersing technology widely, further integrating the world economy, and promoting economic progress in the developing world.** Their ranks will include a growing number based in such countries as China, India, or Brazil. While North America, Japan, and Europe might collectively continue to dominate international political and financial institutions, globalization will take on an increasingly non-Western character. By 2020, globalization could be equated in the popular mind with a rising Asia, replacing its current association with Americanization.

An expanding global economy will increase demand for many raw materials, such as oil. Total energy consumed probably will rise by about 50 percent in the next two decades compared to a 34 percent expansion from 1980-2000, with a greater share provided by petroleum. Most experts assess that with substantial investment in new capacity, overall energy supplies will be sufficient to meet global demands. But on the supply side, many of the areas—the Caspian Sea, Venezuela, and West Africa—that are being counted on to provide increased output involve substantial political or economic risk. Traditional suppliers in the Middle East are also increasingly unstable. **Thus sharper demand-driven competition for resources, perhaps accompanied by a major disruption of oil supplies, is among the key uncertainties**.

- China, India, and other developing countries' growing energy needs suggest a growing preoccupation with energy, shaping their foreign policies.

- For Europe, an increasing preference for natural gas may reinforce regional relationships—such as with Russia or North Africa—given the interdependence of pipeline delivery.

## New Challenges to Governance

**The nation-state will continue to be the dominant unit of the global order, but economic globalization and the dispersion of technologies, especially information technologies, will place enormous new strains on governments**. Growing connectivity will be accompanied by the proliferation of virtual communities of interest, complicating the ability of states to govern. The Internet in particular will spur the creation of even more global movements, which may emerge as a robust force in international affairs.

Part of the pressure on governance will come from new forms of identity politics centered on religious convictions. In a rapidly globalizing world experiencing population shifts, religious identities provide followers with a ready-made community that serves as a "social safety net" in times of need—particularly important to migrants. In particular, **political Islam will have a significant global impact leading to 2020, rallying disparate ethnic and national groups and perhaps even creating an authority that transcends national boundaries.** A combination of factors—youth bulges in many Arab states, poor economic prospects, the influence of religious education, and the Islamization of such institutions as trade unions, nongovernmental organizations, and political parties—will ensure that political Islam remains a major force.

- Outside the Middle East, political Islam will continue to appeal to Muslim migrants who are attracted to the more prosperous West for employment opportunities but do not feel at home in what they perceive as an alien and hostile culture.

Regimes that were able to manage the challenges of the 1990s could be overwhelmed by those of 2020. Contradictory forces will be at work: authoritarian regimes will face new pressures to democratize, but fragile new democracies may lack the adaptive capacity to survive and develop.

**The so-called "third wave" of democratization may be partially reversed by 2020—particularly among the states of the former Soviet Union and in Southeast Asia, some of which never really embraced democracy.** Yet democratization and greater pluralism could gain ground in key Middle Eastern countries which thus far have been excluded from the process by repressive regimes.

With migration on the increase in several places around the world—from North Africa and the Middle East into Europe, Latin America and the Caribbean into the United States, and increasingly from Southeast Asia into the northern regions—more countries will be multi-ethnic and will face the challenge of integrating migrants into their societies while respecting their ethnic and religious identities.

Chinese leaders will face a dilemma over how much to accommodate pluralistic pressures to relax political controls or risk a popular backlash if they do not. Beijing may pursue an "Asian way of democracy," which could involve elections at the local level and a consultative mechanism on the national level, perhaps with the Communist Party retaining control over the central government.

**With the international system itself undergoing profound flux, some of the institutions that are charged with managing global problems may be overwhelmed by them.** Regionally based institutions will be particularly challenged to meet the complex transnational threats posed by terrorism, organized crime, and WMD proliferation. Such post-World War II creations as the United Nations and the international financial institutions risk sliding into obsolescence unless they adjust to the profound changes taking place in the global system, including the rise of new powers.

## Pervasive Insecurity

We foresee a more pervasive sense of insecurity—which may be as much based on psychological perceptions as physical threats—by 2020. **Even as most of the world gets richer, globalization will profoundly shake up the status quo—generating enormous economic, cultural, and consequently political convulsions**. With the gradual integration of China, India, and other emerging countries into the global economy, hundreds of millions of working-age adults will become available for employment in what is evolving into a more integrated world labor market.

- This enormous work force—a growing portion of which will be well educated—will be an attractive, competitive source of low-cost labor at the same time that technological innovation is expanding the range of globally mobile occupations.

- **The transition will not be painless and will hit the middle classes of the developed world in particular**, bringing more rapid job turnover and requiring professional retooling. Outsourcing on a large scale would strengthen the anti-globalization movement. Where these pressures lead will depend on how political leaders respond, how flexible labor markets become, and whether overall economic growth is sufficiently robust to absorb a growing number of displaced workers.

**Weak governments, lagging economies, religious extremism, and youth bulges will align to create a perfect storm for internal conflict in certain regions.** The number of internal conflicts is down significantly since the late 1980s and early 1990s when the breakup of the Soviet Union and Communist regimes in Central Europe allowed suppressed ethnic and nationalistic strife to flare. Although a leveling off point has been reached where we can expect fewer such conflicts than during the last decade, the continued prevalence of troubled and institutionally weak states means that such conflicts will continue to occur.

Some internal conflicts, particularly those that involve ethnic groups straddling national boundaries, risk escalating into regional conflicts. At their most extreme, internal conflicts can result in failing or failed states, with expanses of territory and populations devoid of effective governmental control. Such territories can become sanctuaries for transnational terrorists (such as al-Qa'ida in Afghanistan) or for criminals and drug cartels (such as in Colombia).

**The likelihood of great power conflict escalating into total war in the next 15 years is lower than at any time in the past century, unlike during previous centuries when local conflicts sparked world wars.** The rigidities of alliance systems before World War I and during the interwar period, as well as the two-bloc standoff during the Cold War, virtually assured that small conflicts would be quickly generalized. The growing dependence on global financial and trade networks will help deter interstate conflict but does not eliminate the possibility. Should conflict occur that involved one or more of the great powers, the consequences would be significant. The absence of effective conflict resolution mechanisms in some regions, the rise of nationalism in some states, and the raw emotions and tensions on both sides of some issues—for example, the Taiwan Strait or India/Pakistan issues—could lead to miscalculation. Moreover, advances in modern weaponry—longer ranges, precision delivery, and more destructive conventional munitions—create circumstances encouraging the preemptive use of military force.

Current nuclear weapons states will continue to improve the survivability of their deterrent forces and almost certainly will improve the reliability, accuracy, and lethality of their delivery systems as well as develop capabilities to penetrate missile defenses. The open demonstration of nuclear capabilities by any state would further discredit the current nonproliferation regime, cause a possible shift in the balance of power, and increase the risk of conflicts escalating into nuclear ones. **Countries without nuclear weapons—especially in the Middle East and Northeast Asia—might decide to seek them as it becomes clear that their neighbors and regional rivals are doing so.** Moreover, the assistance of proliferators will reduce the time required for additional countries to develop nuclear weapons.

## Transmuting International Terrorism

**The key factors that spawned international terrorism show no signs of abating over the next 15 years.** Facilitated by global communications, the revival of Muslim identity will create a framework for the spread of radical Islamic ideology inside and outside the Middle East, including Southeast Asia, Central Asia and Western Europe, where religious identity has traditionally not been as strong. This revival has been accompanied by a deepening solidarity among Muslims caught up in national or regional separatist struggles, such as Palestine, Chechnya, Iraq, Kashmir, Mindanao, and southern Thailand, and has emerged in response to government repression, corruption, and ineffectiveness. Informal networks of charitable foundations, *madrassas, hawalas*[1]*,* and other mechanisms will continue to proliferate and be exploited by radical elements; alienation among unemployed youths will swell the ranks of those vulnerable to terrorist recruitment.

**We expect that by 2020 al-Qa'ida will be superceded by similarly inspired Islamic extremist groups**, and there is a substantial risk that broad Islamic movements akin to al-Qa'ida will merge with local separatist movements. Information technology, allowing for instant connectivity, communication, and learning, will enable the terrorist threat to become increasingly decentralized, evolving into an eclectic array of groups, cells, and individuals that do not need a stationary headquarters to plan and carry out operations. Training materials, targeting guidance, weapons know-how, and fund-raising will become virtual (i.e., online).

---

[1] *Hawalas* constitute an informal banking system.

Terrorist attacks will continue to primarily employ conventional weapons, incorporating new twists and constantly adapting to counterterrorist efforts. Terrorists probably will be most original not in the technologies or weapons they use but rather in their operational concepts—i.e., the scope, design, or support arrangements for attacks.

Strong terrorist interest in acquiring chemical, biological, radiological and nuclear weapons increases the risk of a major terrorist attack involving WMD. **Our greatest concern is that terrorists might acquire biological agents or, less likely, a nuclear device, either of which could cause mass casualties.** Bioterrorism appears particularly suited to the smaller, better-informed groups. We also expect that terrorists will attempt cyber attacks to disrupt critical information networks and, even more likely, to cause physical damage to information systems.

## Possible Futures

In this era of great flux, we see several ways in which major global changes could take shape in the next 15 years, from seriously challenging the nation-state system to establishing a more robust and inclusive globalization. In the body of this paper we develop these concepts in four fictional scenarios which were extrapolated from the key trends we discuss in this report. **These scenarios are not meant as actual forecasts**, but they describe possible worlds upon whose threshold we may be entering, depending on how trends interweave and play out:

- *Davos World* provides an illustration of how robust economic growth, led by China and India, over the next 15 years could reshape the globalization process—giving it a more non-Western face and transforming the political playing field as well.

- *Pax Americana* takes a look at how US predominance may survive the radical changes to the global political landscape and serve to fashion a new and inclusive global order.

- *A New Caliphate* provides an example of how a global movement fueled by radical religious identity politics could constitute a challenge to Western norms and values as the foundation of the global system.

- *Cycle of Fear* provides an example of how concerns about proliferation might increase to the point that large-scale intrusive security measures are taken to prevent outbreaks of deadly attacks, possibly introducing an Orwellian world.

Of course, these scenarios illustrate just a few of the possible futures that may develop over the next 15 years, but the wide range of possibilities we can imagine suggests that this period will be characterized by increased flux, particularly in contrast to the relative stasis of the Cold War era. The scenarios are not mutually exclusive: we may see two or three of these scenarios unfold in some combination or a wide range of other scenarios.

## Policy Implications

The role of the United States will be an important shaper of the international order in 2020.  Washington may be increasingly confronted with the challenge of managing—at an acceptable cost to itself—relations with Europe, Asia, the Middle East, and others absent a single overarching threat on which to build consensus.  **Although the challenges ahead will be daunting, the United States will retain enormous advantages, playing a pivotal role across the broad range of issues—economic, technological, political, and military—that no other state will match by 2020.**  Some trends we probably can bank on include dramatically altered alliances and relationships with Europe and Asia, both of which formed the bedrock of US power in the post-World War II period.  The EU, rather than NATO, will increasingly become the primary institution for Europe, and the role which Europeans shape for themselves on the world stage is most likely to be projected through it.  Dealing with the US-Asia relationship may arguably be more challenging for Washington because of the greater flux resulting from the rise of two world-class economic and political giants yet to be fully integrated into the international order.  Where US-Asia relations lead will result as much or more from what the Asians work out among themselves as any action by Washington.  One could envisage a range of possibilities from the US enhancing its role as balancer between contending forces to Washington being seen as increasingly irrelevant.

The US economy will become more vulnerable to fluctuations in the fortunes of others as global commercial networking deepens.  US dependence on foreign oil supplies also makes it more vulnerable as the competition for secure access grows and the risks of supply side disruptions increase.

**While no single country looks within striking distance of rivaling US military power by 2020, more countries will be in a position to make the United States pay a heavy price for any military action they oppose.  The possession of chemical, biological, and/or nuclear weapons** by Iran and North Korea and the possible acquisition of such weapons by others by 2020 **also increase the potential cost of any military action by the US** against them or their allies.

The success of the US-led counterterrorism campaign will hinge on the capabilities and resolve of individual countries to fight terrorism on their own soil.  Counterterrorism efforts in the years ahead—against a more diverse set of terrorists who are connected more by ideology than by geography—will be a more elusive challenge than focusing on a centralized organization such as al-Qa'ida.  **A counterterrorism strategy that approaches the problem on multiple fronts offers the greatest chance of containing—and ultimately reducing—the terrorist threat.**  The development of more open political systems and representation, broader economic opportunities, and empowerment of Muslim reformers would be viewed positively by the broad Muslim communities who do not support the radical agenda of Islamic extremists.

Even if the numbers of extremists dwindle, however, the terrorist threat is likely to remain. The rapid dispersion of biological and other lethal forms of technology increases the potential for an individual not affiliated with any terrorist group to be able to wreak widespread loss of life. Despite likely high-tech breakthroughs that will make it easier to track and detect terrorists at work, the attacker will have an easier job than the defender because the defender must prepare against a large array of possibilities. The United States probably will continue to be called on to help manage such conflicts as Palestine, North Korea, Taiwan, and Kashmir to ensure they do not get out of hand if a peace settlement cannot be reached. However, the scenarios and trends we analyze in the paper suggest the possibility of harnessing the power of the new players in contributing to global security and relieving the US of some of the burden.

**Over the next 15 years the increasing centrality of ethical issues, old and new, have the potential to divide worldwide publics and challenge US leadership.** These issues include the environment and climate change, privacy, cloning and biotechnology, human rights, international law regulating conflict, and the role of multilateral institutions. The United States increasingly will have to battle world public opinion, which has dramatically shifted since the end of the Cold War. Some of the current anti-Americanism is likely to lessen as globalization takes on more of a non-Western face. At the same time, the younger generation of leaders—unlike during the post-World War II period—has no personal recollection of the United States as its "liberator" and is more likely to diverge with Washington's thinking on a range of issues.

In helping to map out the global future, the United States will have many opportunities to extend its advantages, particularly in shaping a new international order that integrates disparate regions and reconciles divergent interests.

# Methodology

To launch the **NIC 2020 Project**, in November 2003 we brought together some 25 leading outside experts from a wide variety of disciplines and backgrounds to engage in a broad-gauged discussion with Intelligence Community analysts. We invited three leading "futurists"—Ted Gordon of the UN's Millennium Project; Jim Dewar, Director of the RAND Corporation's Center for Longer Range Global Policy and the Future of the Human Condition; and Ged Davis, former head of Shell International's scenarios project[2]—to discuss their most recent work and the methodologies they employed to think about the future. Princeton University historian Harold James gave the keynote address, offering lessons from prior periods of "globalization."

We surveyed and studied various methodologies (see box on page 22) and reviewed a number of recent "futures" studies. Besides convening a meeting of counterparts in the UK, Canada, Australia, and New Zealand to learn their thinking, we organized six regional conferences in countries on four continents—one in the United Kingdom, South Africa, Singapore, and Chile, two in Hungary—to solicit the views of foreign experts from a variety of backgrounds—academics, business people, government officials, members of nongovernmental organizations and other institutions—who could speak authoritatively on the key drivers of change and conceptualize broad regional themes. Our regional experts also contributed valuable insights on how the rest of the world views the United States. In addition to the conferences held overseas, which included hundreds of foreign participants, we held a conference in the Washington, DC area on India.

We augmented these discussions with conferences and workshops that took a more in-depth view of specific issues of interest, including new technologies, the changing nature of warfare, identity politics, gender issues, climate change and many others (see box on page 20 for a complete list of the conferences). Participants explored key trends that were presented by experts and then developed alternative scenarios for how the trends might play out over the next 15 years. And we consulted numerous organizations and individuals on the substantive aspects of this study, as well as on methodologies and approaches for thinking about the future.

- The UN Millennium Project—an independent body that advises the UN on strategies for achieving the Millennium development goals—provided invaluable data on cross-cutting issues. We also consulted the *Eurasia Group, Oxford Analytica, CENTRA Technologies*, and the *Stimson Center*.

- Other individual scholars we consulted included Michael F. Oppenheimer, President, Global Scenarios, who facilitated several of our sessions and informed our thinking on methodologies; Georgetown and now Princeton Professor John Ikenberry, who organized several seminars of academic experts over the course of more than a

---

[2] Shell International Limited has for decades used scenarios to identify business risks and opportunities. Ged Davis led this effort for many years.

year to examine various aspects of US preeminence and critique preliminary drafts of the report; Enid Schoettle, who was one of the architects of *Global Trends 2015*; Professor Barry B. Hughes, Graduate School of International Studies, University of Denver, whose related statistical and scenario work is featured on our Web site; Anne Solomon, Senior Adviser on Technology Policy and Director of the Biotechnology and Public Policy Program at the Center for Strategic and International Studies in Washington, DC, who organized several stimulating conferences on S&T topics; Elke Matthews, an independent contractor who conducted substantial open-source research; Philip Jenkins, Distinguished Professor of History and Religious Studies, Pennsylvania State University, who provided invaluable insights on global trends pertaining to religion; Nicholas Eberstadt, Henry Wendt Chair in Political Economy, American Enterprise Institute, who provided us with important perspectives on demographic issues; and Jeffrey Herbst, Chair, Department of Politics, Princeton University, who was instrumental in our analysis of issues pertaining to Africa.

---

**NIC 2020 Project Conferences and Workshops**

Presentation by Joint Doctrine and Concepts Center (MoD/UK)—CIA Headquarters (September 2003)
Conference on Anti-Americanism—Wye Plantation (October 2003)
Inaugural NIC 2020 Project Conference—Washington, DC (November 2003)
Professor Ikenberry's series of International Relations Roundtables—Georgetown University (November 2003-November 2004)
Joint US-Commonwealth Intelligence Officials' Conference —Washington, DC (December 2003)
African Experts' Roundtable—Washington, DC (January 2004)
Middle East NIC 2020 Workshop—Wilton Park, UK (March 2004)
Africa NIC 2020 Workshop—Johannesburg, South Africa (March 2004)
Global Evolution of Dual-Use Biotechnology—Washington, DC (March 2004)
Russia and Eurasia NIC 2020 Workshop—Budapest, Hungary (April 2004)
Europe NIC 2020 Workshop—Budapest, Hungary (April 2004)
Global Identity Roundtable Discussion—CIA Headquarters (May 2004)
Asia NIC 2020 Workshop—Singapore (May 2004)
Conference on The Changing Nature of Warfare—Center for Naval Analysis (May 2004)
Latin America NIC 2020 Workshop—Santiago, Chile (June 2004)
Technological Frontiers, Global Power, Wealth, and Conflict—Center for Strategic and International Studies (CSIS) (June 2004)
Climate Change—University of Maryland (June 2004)
NSA Tech 2020—Baltimore, Maryland (June 2004)
Conference on Muslims in Europe—Oxford, England (July 2004)
Women in 2020—Washington, DC (August 2004)
Business Leader Roundtable Discussion—CIA Headquarters (September 2004)
India and Geopolitics in 2020–Rosslyn, Virginia (September 2004)
Stimson Center-sponsored roundtables on Scenarios—Washington, DC (Spring-Summer, 2004)
Information and Communications, Technological and Social Cohesion and the Nation-State—Washington, DC (September 2004)
Wrap-Up NIC 2020 Project Workshop—Virginia (October 2004)
Consultation on Preliminary NIC 2020 Draft with UK experts and the International Institute of Strategic Studies—London, England (October 2004)

- The following organizations arranged the regional conferences for the project: Wilton Park, Central European University, Bard College, the South African Institute for International Affairs, Adolfo Ibañez University, Nueva Mayoria, and the Asia Society. Timothy Sharp and Professor Ewan Anderson of Sharp Global Solutions Ltd arranged a conference in London of UK experts to critique a preliminary draft of the report.

- We also want to thank our colleagues in the US Intelligence Community, who provided us with useful data and shared their ideas about global trends.

**Scenario Development Process**

While straight-line projections are useful in establishing a baseline and positing a mainline scenario, they typically present a one-dimensional view of how the future might unfold and tend to focus attention exclusively on the "prediction." Scenarios offer a more dynamic view of possible futures and focus attention on the underlying interactions that may have particular policy significance. They are especially useful in thinking about the future during times of great uncertainty, which we believe is the case for the next 15 years. Scenarios help decisionmakers to break through conventional thinking and basic assumptions so that a broader range of possibilities can be considered—including new risks and opportunities.

The six international workshops generated an enormous amount of data and analysis on the key drivers that are likely to lead to regional change in the 2020 timeframe. The NIC 2020 Project staff conducted additional research, drafted papers, and initiated follow-up roundtable discussions and conferences. We analyzed the findings from the regional workshops, highlighted key regional trends that had global implications, and looked at the regional product in its totality to identify salient cross-regional trends. These key findings were set aside as the raw material for development of the global scenarios.

To jumpstart the global scenario development process, the NIC 2020 Project staff created a Scenario Steering Group (SSG)—a small aggregation of respected members of the policy community, think tanks, and analysts from within the Intelligence Community—to examine summaries of the data collected and consider scenario concepts that take into account the interaction between key drivers of global change. SSG examined the product of the international workshops and explored fledgling scenarios for plausibility and policy relevance.

We studied extensively key futures work developed in the public and private sectors that employed scenario techniques, identified the "best practices," and then developed our own unique approach, combining trend analysis and scenarios. Papers that influenced our work include those produced by Goldman Sachs, the UK Ministry of Defense, and Shell International, Ltd. (see box on page 22).

## Scenario and Futures Work That Influenced Our Thinking

Our consultations with Ged Davis, formerly the leader of **Shell International's** scenario-building effort, affirmed our intent to develop scenarios for policymakers. **Shell builds global scenarios every three years to help its leaders make better decisions. Following initial research, Shell's team spends about a year conducting interviews and holding workshops to develop and finalize the scenarios, seeking throughout the process to ensure a balance between unconventional thinking and plausibility. We used a similar approach. We also benefited from consultations with other organizations that do futures work:**

**The Joint Doctrine and Concepts Centre**, an integral part of the UK Ministry of Defense, undertook an ambitious attempt to develop a coherent view of how the world might develop over the next 30 years in ways that could alter the UK's security. The project—Strategic Trends—was designed to assist the MOD in gaining a strategic understanding of future threats, risks, challenges, and opportunities.

Meta-Analysis of Published Material on Drivers and Trends, produced by the **UK Defense Evaluation and Research Agency,** reviewed over 50 futures studies.

**The RAND Corporation**—as part of a parallel, NIC-sponsored effort to update its 2001 monograph *The Global Revolution: Bio/Nano/Materials Trends and Their Synergies with IT by 2015*—provided substantive guidance by delineating technology trends and their interaction; identifying applications that will transform the future; commenting extensively on drafts; and providing thought-provoking, technology-driven scenario concepts.

**Peter Schwartz, Chairman, Global Business Network and author of Inevitable Surprises,** provided us with invaluable insights on the nature of surprise, including the use of drivers, the interpretation of insights across disciplines, and the application of scenario work to the private sector.

**Toffler Associates** contributed ideas at several points, including in association with the NSA Tech 2020 project (see below). In addition, Drs. Alvin and Heidi Toffler participated in our capstone conference, sharing their insights on understanding the future based on their vast experience in the field.

The **National Security Agency's** project—**Tech 2020**—also helped identify key technology convergences expected to impact society between now and 2020. We have incorporated valuable insights from this project and are grateful to NSA for stimulating a rewarding Intelligence Community dialogue on future trends.

After scenario concepts were explored, critiqued, and debated within the SSG and with other groups that the NIC engaged, eight global scenarios that held particular promise were developed. The NIC then held a wrap-up workshop with a broader group of experts to examine the eight scenarios, discuss the merits and weaknesses of each, and ultimately narrow the number of scenarios included in the final publication to four. The scenarios depicted in this publication were selected for their relevance to policymakers and because they cause us to question key assumptions about the future—but they do not attempt to predict it. Nor are they mutually exclusive.

## Interactive Tools
Significantly, the NIC 2020 Project also employs information technology and analytic tools unavailable in earlier NIC efforts. Its global sweep and scope required that we engage in a continuing, worldwide dialogue about the future. With the help of CENTRA Technologies, we created an interactive, password-protected Web site to serve as a repository for discussion papers and workshop summaries. The site also provided a link to massive quantities of basic data for reference and analysis. It contained interactive tools to keep our foreign and domestic experts engaged and created "hands-on" computer simulations that allowed novice and expert alike to develop their own scenarios.[3] Much of this supporting material involving the Empirical Web-boxes Scenario capability has now been transferred to the open, unclassified NIC Web site with publication of this report.

---

[3] To access these new innovations log on to the NIC website: www.cia.gov/nic.

# Introduction

The international order is in the midst of profound change: at no time since the formation of the Western alliance system in 1949 have the shape and nature of international alignments been in such a state of flux as they have during the past decade. As a result, the world of 2020 will differ markedly from the world of 2004, and in the intervening years the United States will face major international challenges that differ significantly from those we face today. The very magnitude and speed of change resulting from a globalizing world—regardless of its precise character—will be a defining feature of the world out to 2020. Other significant characteristics include:

- The contradictions of globalization.

- Rising powers: the changing geopolitical landscape.

- New challenges to governance.

- A more pervasive sense of insecurity.

As with previous upheavals, the seeds of major change have been laid in the trends apparent today. Underlying the broad characteristics listed above are a number of specific trends that overlap and play off each other:

- The expanding global economy.

- The accelerating pace of scientific change and the dispersion of dual-use technologies.

- Lingering social inequalities.

- Emerging powers.

- The global aging phenomenon.

- Halting democratization.

- A spreading radical Islamic ideology.

- The potential for catastrophic terrorism.

- The proliferation of weapons of mass destruction.

- Increased pressures on international institutions.

As we survey the next 15 years, the role of the United States will be an important variable in how the world is shaped, influencing the path that states and nonstate actors choose to follow. In addition to the pivotal role of the United States, international bodies including international organizations, multinational corporations, nongovernmental organizations (NGOs) and others can mitigate distinctly negative trends, such as greater insecurity, and advance positive trends.

## Mapping the Global Future

How we mentally map the world will be different in 2020. Traditional geographic groupings will increasingly lose salience in international relations. Since the end of the Cold War, scholars have questioned the utility of the East vs. West concept that emerged in the late 1940s as an intellectual justification for American engagement in Europe. Eurasia as a concept supplanting the former Soviet Union seems irrelevant as many former Soviet members go their own way and the prospect of Moscow reasserting control seems improbable. The usefulness of the West as a concept has been questioned by the growing philosophic divisions between the US and Western Europe over sovereignty and multilateralism and the increasing power on the world scene of traditionally non-Western powers.

As with the East-West divide, the traditional North-South faultline may not be a meaningful concept for the world in 2020 owing to globalization and the expected rise of China and India, which have been considered part of the "South" because of their level of development. Traditional issues of North-South inequalities, trade, and assistance will certainly keep coming to the fore, but some high-growth developing countries, especially China and India, probably will be among the economic heavy-weights or "haves." They will not be "Western" in the traditional sense but also may not be seen as representative of the underdeveloped or still developing countries. China, in particular, may see itself as having been restored as a great power after several centuries of decline.

Divisions other than economic may also shape how we view the world. We anticipate that religion will play an increasing role in how many people define their identity. For many societies, divisions between and within religious groups may become boundaries as significant as national borders. We particularly see Christian-Muslim divides in Southeast Asia, splits within the Muslim world between Shia and Sunni communities, and islands of potential religious or ethnic disaffection in Europe, Russia, and China as figuring prominently in the 2020 geographic outlook.

One current concept that may still hold true in 2020 is that of an increasing arc of instability ranging from Southeast Asia, where the possibility exists of growing radical Islam and terrorism, to Central Asia, where we see the possibility of "failed," non-democratic states. The arc includes many Middle Eastern and African countries, some of whom may have fallen further behind or have only just begun to connect with the global economy. Globalization above all will have replaced the former divide among the industrialized West; Communist East; and the developing, non-aligned, or Third World. New alignments instead will be between those countries, or even parts of countries or hubs, that are integrating into a global community and those that are not integrating for economic, political or social reasons. For those mega-cities or hubs that are the engine behind globalization, the financial and telecommunications links they forge with each other may matter as much or more than national boundaries.

# The Contradictions of Globalization

Whereas in **Global Trends 2015** we viewed globalization—growing interconnectedness reflected in the expanded flows of information, technology, capital, goods, services, and people throughout the world—as among an array of key drivers, we now view it more as a "mega-trend"—a force so ubiquitous that it will substantially shape all of the other major trends in the world of 2020.

*"[By 2020] globalization is likely to take on much more of a 'non-Western' face…"*

The reach of globalization was substantially broadened during the last 20 years by Chinese and Indian economic liberalization, the collapse of the Soviet Union, and the worldwide information technology revolution. Through the next 15 years, it will sustain world economic growth, raise world living standards, and substantially deepen global interdependence. At the same time, it will profoundly shake up the status quo almost everywhere—generating enormous economic, cultural, and consequently political convulsions.

Certain aspects of globalization, such as the growing global inter-connectedness stemming from the information technology revolution, are likely to be irreversible. Real-time communication, which has transformed politics almost everywhere, is a phenomenon that even repressive governments would find difficult to expunge.

- It will be difficult, too, to turn off the phenomenon of entrenched economic interdependence, although the pace of global economic expansion may ebb and flow. Interdependence has widened the effective reach of multinational business, enabling smaller firms as well as large multinationals to market across borders and bringing heretofore non-traded services into the international arena.

Yet the process of globalization, powerful as it is, could be substantially slowed or even reversed, just as the era of globalization in the late 19th and early 20th centuries was reversed by catastrophic war and global depression. Some features that we associate with the globalization of the 1990s—such as economic and political liberalization—are prone to "fits and starts" and probably will depend on progress in multilateral negotiations, improvements in national governance, and the reduction of conflicts. The freer flow of people across national borders will continue to face social and political obstacles even when there is a pressing need for migrant workers.

*"India and China probably will be among the economic heavyweights or 'haves.'"*

27

## What Would An Asian Face on Globalization Look Like?

Rising Asia will continue to reshape globalization, giving it less of a "Made in the USA" character and more of an Asian look and feel. At the same time, Asia will alter the rules of the globalizing process. By having the fastest-growing consumer markets, more firms becoming world-class multinationals, and greater S&T stature, Asia looks set to displace Western countries as the focus for international economic dynamism—provided Asia's rapid economic growth continues.

Asian finance ministers have considered establishing an Asian monetary fund that would operate along different lines from IMF, attaching fewer strings on currency swaps and giving Asian decision-makers more leeway from the "Washington macro-economic consensus."

- In terms of capital flows, rising Asia may still accumulate large currency reserves—currently $850 billion in Japan, $500 billion in China, $190 billion in Korea, and $120 million in India, or collectively three-quarters of global reserves—but the percentage held in dollars will fall. A basket of reserve currencies including the yen, renminbi, and possibly rupee probably will become standard practice.

- Interest-rate decisions taken by Asian central bankers will impact other global financial markets, including New York and London, and the returns from Asian stock markets are likely to become an increasing global benchmark for portfolio managers.

As governments devote more resources to basic research and development, rising Asia will continue to attract applied technology from around the world, including cutting-edge technology, which should boost their high performance sectors. We already anticipate (as stated in the text) that the Asian giants may use the power of their markets to set industry standards, rather than adopting those promoted by Western nations or international standards bodies. The international intellectual property rights regime will be profoundly molded by IPR regulatory and law enforcement practices in East and South Asia.

Increased labor force participation in the global economy, especially by China, India, and Indonesia, will have enormous effects, possibly spurring internal and regional migrations. Either way it will have a large impact, determining the relative size of the world's greatest new "mega-cities" and, perhaps, act as a key variable for political stability/instability for decades to come. To the degree that these vast internal migrations spill over national borders—currently, only a miniscule fraction of China's 100 million internal migrants end up abroad—they could have major repercussions for other regions, including Europe and North America.

An expanded Asian-centric cultural identity may be the most profound effect of a rising Asia. Asians have already begun to reduce the percentage of students who travel to Europe and North America with Japan and—most striking—China becoming educational magnets. A new, more Asian cultural identity is likely to be rapidly packaged and distributed as incomes rise and communications networks spread. Korean pop singers are already the rage in Japan, Japanese *anime* have many fans in China, and Chinese kung-fu movies and Bollywood song-and-dance epics are viewed throughout Asia. Even Hollywood has begun to reflect these Asian influences—an effect that is likely to accelerate through 2020.

Moreover, the character of globalization probably will change just as capitalism changed over the course of the 19<sup>th</sup> and 20<sup>th</sup> centuries. While today's most advanced nations—especially the United States—will remain important forces driving capital, technology and goods, globalization is likely to take on much more of a "non-Western face" over the next 15 years.

- Most of the increase in world population and consumer demand through 2020 will take place in today's developing nations—especially China, India, and Indonesia—and multinational companies from today's advanced nations will adapt their "profiles" and business practices to the demands of these cultures.

- Able to disperse technology widely and promote economic progress in the developing world, corporations already are seeking to be "good citizens" by allowing the retention of non-Western practices in the workplaces in which they operate. Corporations are in the position to make globalization more palatable to people concerned about preserving unique cultures.

- New or expanding corporations from countries lifted up by globalization will make their presence felt globally through trade and investments abroad.

- Countries that have benefited and are now in position to weigh in will seek more power in international bodies and greater influence on the "rules of the game."

- In our interactions, many foreign experts have noted that while popular opinion in their countries favors the material benefits of globalization, citizens are opposed to its perceived "Americanization," which they see as threatening to their cultural and religious values. The conflation of globalization with US values has in turn fueled anti-Americanism in some parts of the world.

*"…the world economy is projected to be about 80 percent larger in 2020 than it was in 2000, and average per capita income to be roughly 50 percent higher."*

Currently, about two-thirds of the world's population live in countries that are connected to the global economy. Even by 2020, however, the benefits of globalization won't be global. Over the next 15 years, gaps will widen between those countries benefiting from globalization—economically, technologically, and socially—and those underdeveloped nations or pockets within nations that are left behind. Indeed, we see the next 15 years as a period in which the perceptions of the contradictions and uncertainties of a globalized world come even more to the fore than is the case today.

### An Expanding and Integrating Global Economy

The world economy is projected to be about 80 percent larger in 2020 than it was in 2000 and average per capita income to be roughly 50 percent higher. Large parts of the world will enjoy unprecedented prosperity, and a *numerically* large middle class will be created for the first time in some formerly poor countries. The social structures in

**What Could Derail Globalization?**

The process of globalization, powerful as it is, could be substantially slowed or even stopped. Short of a major global conflict, which we regard as improbable, another large-scale development that we believe could *stop* globalization would be a pandemic. However, other catastrophic developments, such as terrorist attacks, could *slow* its speed.

Some experts believe it is only a matter of time before a new *pandemic* appears, such as the 1918–1919 influenza virus that killed an estimated 20 million worldwide. Such a pandemic in megacities of the developing world with poor health-care systems—in Sub-Saharan Africa, China, India, Bangladesh or Pakistan—would be devastating and could spread rapidly throughout the world. Globalization would be endangered if the death toll rose into the millions in several major countries and the spread of the disease put a halt to global travel and trade during an extended period, prompting governments to expend enormous resources on overwhelmed health sectors. On the positive side of the ledger, the response to SARS showed that international surveillance and control mechanisms are becoming more adept at containing diseases, and new developments in biotechnologies hold the promise of continued improvement.

A slow-down could result from *a pervasive sense of economic and physical insecurity* that led governments to put controls on the flow of capital, goods, people, and technology that stalled economic growth. Such a situation could come about in response to terrorist attacks killing tens or even hundreds of thousands in several US cities or in Europe or to widespread cyber attacks on information technology. Border controls and restrictions on technology exchanges would increase economic transaction costs and hinder innovation and economic growth. Other developments that could stimulate similar restrictive policies include a popular backlash against globalization prompted, perhaps, by white collar rejection of outsourcing in the wealthy countries and/or resistance in poor countries whose peoples saw themselves as victims of globalization.

those developing countries will be transformed as growth creates a greater middle class. Over a long time frame, there is the potential, so long as the expansion continues, for more traditionally poor countries to be pulled closer into the globalization circle.

Most forecasts to 2020 and beyond continue to show higher annual growth for developing countries than for high-income ones. Countries such as China and India will be in a position to achieve higher economic growth than Europe and Japan, whose aging work forces may inhibit their growth. Given its enormous population—and assuming a reasonable degree of real currency appreciation—the dollar value of China's gross national product (GNP) may be the second largest in the world by 2020. For similar reasons, the value of India's output could match that of a large European country. The economies of other developing countries,

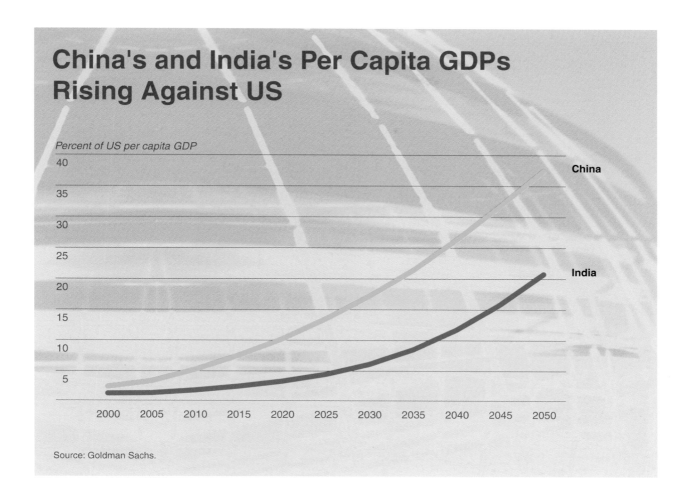

# China's and India's Per Capita GDPs Rising Against US

Percent of US per capita GDP

China

India

2000  2005  2010  2015  2020  2025  2030  2035  2040  2045  2050

Source: Goldman Sachs.

such as Brazil and Indonesia, could surpass all but the largest European economies by 2020.[4]

- Even with all their dynamic growth, Asia's "giants" and others are not likely to compare qualitatively to the economies of the US or even some of the other rich countries. They will have some dynamic, world-class sectors, but more of their populations will work on farms, their capital stocks will be less sophisticated, and their financial systems are likely to be less efficient than those of other wealthy countries.

---

[4] *Dreaming with the BRICS*, Goldman Sachs study, October 2003.

***Continued Economic Turbulence.***
Sustained high-growth rates have historical precedents. China already has had about two decades of 7 percent and higher growth rates, and Japan, South Korea, and Taiwan have managed in the past to achieve annual rates averaging around 10 percent for a long period.

Fast-developing countries have historically suffered sudden setbacks, however, and economic turbulence is increasingly likely to spill over and upset broader international relations. Many emerging markets—such as Mexico in the mid-1990s and Asian countries in the late 1990s—suffered negative effects from the abrupt reversals of capital movements, and China and India may

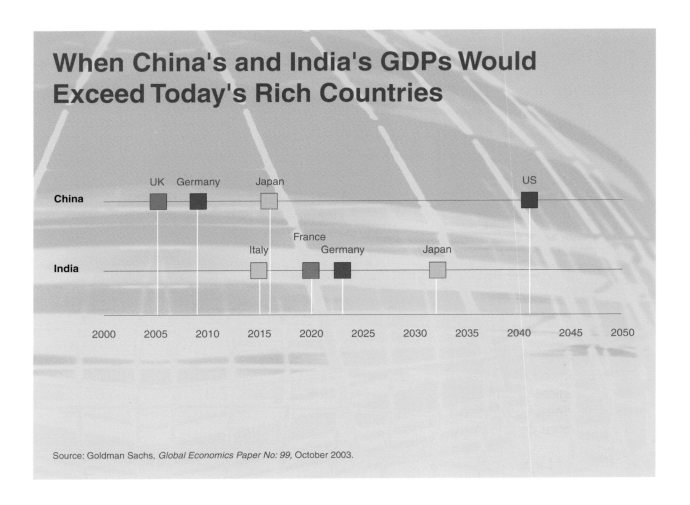

## When China's and India's GDPs Would Exceed Today's Rich Countries

China

UK    Germany        Japan                                    US

India

Italy        France    Germany              Japan

2000   2005   2010   2015   2020   2025   2030   2035   2040   2045   2050

Source: Goldman Sachs, *Global Economics Paper No: 99,* October 2003.

encounter similar problems. The scale of the potential reversals would be unprecedented, and it is unclear whether current international financial mechanisms would be in a position to forestall wider economic disruption.

*"Competitive pressures will force companies based in the advanced economies to 'outsource' many blue- and white-collar jobs."*

With the gradual integration of China, India, and other developing countries into the global economy, hundreds of millions of working-age adults will join what is becoming, through trade and investment flows, a more interrelated world labor market. World patterns of production, trade, employment, and wages will be transformed.

- This enormous work force—a growing portion of which will be well educated—will be an attractive, competitive source of low-cost labor at the same time that technological innovation is expanding the range of globally mobile occupations.

- Competition from these workers will increase job "churning," necessitate professional retooling, and restrain wage growth in some occupations.

Where these labor market pressures lead will depend on how political leaders and policymakers respond. Against the backdrop of a global economic recession, such resources could unleash widespread protectionist sentiments. As long as sufficiently robust economic growth and labor market flexibility are sustained, however, intense international competition is unlikely to cause *net* job "loss" in the advanced economies.

- The large number of new service sector jobs that will be created in India and elsewhere in the developing world, for example, will likely exceed the supply of workers with those specific skills in the advanced economies.

- Job turnover in advanced economies will continue to be driven more by technological change and the vicissitudes of domestic rather than international competition.

***Mobility and Laggards.*** Although the living standards of many people in developing and underdeveloped countries will rise over the next 15 years, per capita incomes in most countries will not compare to those of Western nations by 2020. There will continue to be large numbers of poor even in the rapidly emerging economies, and the proportion of those in the middle stratum is likely to be significantly less than is the case for today's developed nations. Experts estimate it could take China another 30 years beyond 2020 for per capita incomes to reach *current* rates in developed economies.

- Even if, as one study estimates, China's middle class could make up as much as 40 percent of its population by 2020—double what it is

now—it would be still well below the 60 percent level for the US. And per capita income for China's middle class would be substantially less than equivalents in the West.

- In India, there are now estimated to be some 300 million middle-income earners making $2,000-$4,000 a year. Both the number of middle earners and their income levels are likely to rise rapidly, but their incomes will continue to be substantially below averages in the US and other rich countries even by 2020.

- However, a $3,000 annual income is considered sufficient to spur car purchases in Asia; thus rapidly rising income levels for a ***growing middle class*** will combine to mean a huge consumption explosion, which is already evident.

Widening income and regional disparities will not be incompatible with a growing middle class and increasing overall wealth. In India, although much of the west and south may have a large middle class by 2020, a number of regions such as Bihar, Uttar Pradesh, and Orissa will remain underdeveloped.

Moreover, countries not connected to the world economy will continue to suffer. Even the most optimistic forecasts admit that economic growth fueled by globalization will leave many countries in poverty over the next 15 years.

- Scenarios developed by the World Bank indicate, for example, that Sub-Saharan Africa will be far behind even under the most optimistic scenario. The region currently has the largest share of people living on less than $1 per day.

If the growing problem of abject poverty and bad governance in troubled states in Sub-Saharan Africa, Eurasia, the Middle East, and Latin America persists, these areas will become more fertile grounds for terrorism, organized crime, and pandemic disease. Forced migration also is likely to be an important dimension of any downward spiral. The international community is likely to face choices about whether, how, and at what cost to intervene.

*"...the greatest benefits of globalization will accrue to countries and groups that can access and adopt new technologies."*

**The Technology Revolution**
The trend toward rapid, global diffusion of technology will continue, although the stepped-up technology revolution will not benefit everyone equally.

- Among the drivers of the growing availability of technology will be the growing two-way flow of high-tech brain power between developing countries and Western countries, the increasing size of the technologically literate workforce in some developing countries, and efforts by multinational corporations to diversify their high-tech operations.

New technology applications will foster dramatic improvements in human knowledge and individual well-being. Such benefits include medical breakthroughs that begin to cure or

mitigate some common diseases and stretch lifespans, applications that improve food and potable water production, and expansion of wireless communications and language translation technologies that will facilitate transnational business, commercial, and even social and political relationships.

Moreover, future technology trends will be marked not only by accelerating advancements in individual technologies but also by a force-multiplying convergence of the technologies—information, biological, materials, and nanotechnologies—that have the potential to revolutionize all dimensions of life. Materials enabled with nanotechnology's sensors and facilitated by information technology will produce myriad devices that will enhance health and alter business practices and models. Such materials will provide new knowledge about environment, improve security, and reduce privacy. Such interactions of these technology trends—coupled with agile manufacturing methods and equipment as well as energy, water, and transportation technologies—will help China's and India's prospects for joining the "First World." Both countries are investing in basic research in these fields and are well placed to be leaders in a number of key fields. Europe risks slipping behind Asia in creating some of these technologies. The United States is still in a position to retain its overall lead, although it must increasingly compete with Asia and may lose significant ground in some sectors.

***To Adaptive Nations Go Technology 's Spoils.*** The gulf between "haves" and "have-nots" may widen as the greatest benefits of globalization accrue to countries and groups that can access and adopt new technologies. Indeed, a nation's level of technological achievement generally will be defined in terms of its investment in ***integrating and applying*** the new, globally available technologies—whether the technologies are acquired through a country's own basic research or from technology leaders. Nations that remain behind in adopting technologies are likely to be those that have failed to pursue policies that support application of new technologies—such as good governance, universal education, and market reforms—and not solely because they are poor.

Those that employ such policies can leapfrog stages of development, skipping over phases that other high-tech leaders such as the United States and Europe had to traverse in order to advance. China and India are well positioned to achieve such breakthroughs. Yet, even the poorest countries will be able to leverage prolific, cheap technologies to fuel—although at a slower rate—their own development.

- As nations like China and India surge forward in funding critical science and engineering education, research, and other infrastructure investments, they will make considerable strides in manufacturing and marketing a full range of technology applications— from software and pharmaceuticals to wireless sensors and smart-materials products.

Rapid technological advances outside the United States could enable other countries to set the rules for design, standards, and implementation, and for molding privacy, information security, and intellectual property rights (IPR).

- Indeed, international IPR enforcement is on course for dramatic change. Countries like China and India will, because of the purchasing power of their huge markets, be able to shape the implementation of some technologies and step on the intellectual property rights of others. The attractiveness of these large markets will tempt multinational firms to overlook IPR indiscretions that only minimally affect their bottom lines. Additionally, as many of the expected advancements in technology are anticipated to be in medicine, there will be increasing pressure from a humanitarian and moral perspective to "release" the property rights "for the good of mankind."

Nations also will face serious challenges in oversight, control, and prohibition of sensitive technologies. With the same technology, such as sensors, computing, communication, and materials, increasingly being developed for a range of applications in both everyday, commercial settings and in critical military applications the monitoring and control of the export of technological components will become more difficult. Moreover, joint ventures, globalized markets and the growing proportion of private sector capital in basic R&D will undermine nation-state efforts to keep tabs on sensitive technologies.

- Questions concerning a country's ethical practices in the technology realm—such as with genetically modified foods, data privacy, biological material research, concealable sensors, and biometric devices—may become an increasingly important factor in international trade policy and foreign relations.

## Biotechnology:  Panacea and Weapon

The biotechnological revolution is at a relatively early stage, and major advances in the biological sciences coupled with information technology will continue to punctuate the 21st century.  Research will continue to foster important discoveries in innovative medical and public health technologies, environmental remediation, agriculture, biodefense, and related fields.

On the positive side, biotechnology could be a "leveling" agent between developed and developing nations, spreading dramatic economic and healthcare enhancements to the neediest areas of the world.

- Possible breakthroughs in biomedicine such as an antiviral barrier will reduce the spread of HIV/AIDS, helping to resolve the ongoing humanitarian crisis in Sub-Saharan Africa and diminishing the potentially serious drag on economic growth in developing countries like India and China.  Biotechnology research and innovations derived from continued US investments in Homeland Security—such as new therapies that might block a pathogen's ability to enter the body—may eventually have revolutionary healthcare applications that extend beyond protecting the US from a terrorist attack.

- More developing countries probably will invest in indigenous biotechnology developments, while competitive market pressures increasingly will induce firms and research institutions to seek technically capable partners in developing countries.

However, even as the dispersion of biotechnology promises a means of improving the quality of life, it also poses a major security concern.  As biotechnology information becomes more widely available, the number of people who can potentially misuse such information and wreak widespread loss of life will increase.  An attacker would appear to have an easier job—because of the large array of possibilities available—than the defender, who must prepare against them all.  Moreover, as biotechnology advances become more ubiquitous, stopping the progress of offensive BW programs will become increasingly difficult.  Over the next 10 to 20 years there is a risk that advances in biotechnology will augment not only defensive measures but also offensive biological warfare (BW) agent development and allow the creation of advanced biological agents designed to target specific systems—human, animal, or crop.

Lastly, some biotechnology techniques that may facilitate major improvements in health also will spur serious ethical and privacy concerns over such matters as comprehensive genetic profiling; stem cell research; and the possibility of discovering DNA signatures that indicate predisposition for disease, certain cognitive abilities, or anti-social behavior.

At the same time, technology will be a source of tension in 2020: from competition over creating and attracting the most critical component of technological advancement—people—to resistance among some cultural or political groups to the perceived privacy-robbing or homogenizing effects of pervasive technology.

## Lingering Social Inequalities

Even with the potential for technological breakthroughs and the dispersion of new technologies, which could help reduce inequalities, significant social welfare disparities within the developing and between developing and OECD countries will remain until 2020.

Over the next 15 years, illiteracy rates of people 15 years and older will fall, according to UNESCO, but they will still be 17 times higher in poor and developing countries than those in OECD[5] countries. Moreover, illiteracy rates among women will be almost twice as high as those among men. Between 1950 and 1980 life expectancy between the more- and less-developed nations began to converge markedly; this probably will continue to be the case for many developing countries, including the most populous. However, by US Census Bureau projections, over 40 countries—including many African countries, Central Asian states, and Russia—are projected to have a lower life expectancy in 2010 than they did in 1990.

Even if effective HIV/AIDS prevention measures are adopted in various countries, the social and economic impact of the millions already infected with the disease will play out over the next 15 years.

- The rapid rise in adult deaths caused by AIDS has left an unprecedented number of orphans in Africa. Today in some African countries one in ten children is an orphan, and the situation is certain to worsen.

The debilitation and death of millions of people resulting from the AIDS pandemic will have a growing impact on the economies of the hardest-hit countries, particularly those in Sub-Saharan Africa, where more than 20 million are believed to have died from HIV/AIDS since the early 1980s. Studies show that household incomes drop by 50 to 80 percent when key earners become infected. In "second wave" HIV/AIDS countries—Nigeria, Ethiopia, Russia, India, China, Brazil, Ukraine, and the Central Asian states—the disease will continue to spread beyond traditional high-risk groups into the general population. As HIV/AIDS spreads, it has the potential to derail the economic prospects of many up-and-coming economic powers.

---

[5] The OECD, Organization for Economic Cooperation and Development, an outgrowth of the Marshall Plan-era Organization for European Economic Cooperation, boasts 30 members from among developed and emerging-market nations and active relationships with 70 others around the world.

## The Status of Women in 2020

By 2020, women will have gained more rights and freedoms—in terms of education, political participation, and work force equality—in most parts of the world, but UN and World Health Organization data suggest that the gender gap will not have been closed even in the developed countries and still will be wide in developing regions. Although women's share in the global work force will continue to rise, wage gaps and regional disparities will persist.

- Although the difference between women's and men's earnings narrowed during the past 10 years, women continue to receive less pay than men. For example, a UN study in 2002 showed that in 27 of 39 countries surveyed—both in OECD and developing countries—women's wages were 20 to 50 percent less than men's for work in manufacturing.

Certain factors will tend to work against gender equality while others will have a positive impact.

### *Factors Impeding Equality*

In regions where high **youth bulges** intersect with historical patterns of patriarchal bias, the added pressure on infrastructure will mean intensified competition for limited public resources and an increased probability that females will not receive equal treatment. For instance, if schools cannot educate all, boys are likely to be given first priority. Yet views are changing among the younger generation. In the Middle East, for example, many younger Muslims recognize the importance of educated wives as potential contributors to family income.

In countries such as China and India, where there is a pervasive "son preference" reinforced by government **population control policies**, women face increased risk not only of female infanticide but also of kidnapping and smuggling from surrounding regions for the disproportionately greater number of unattached males. Thus far, the preference for male children in China has led to an estimated shortfall of 30 million women.

Such statistics suggest that the global **female trafficking industry**, which already earns an estimated $4 billion every year, is likely to expand, making it the second most profitable criminal activity behind global drug trafficking.

The feminization of **HIV/AIDS** is another worrisome trend. Findings from the July 2004 Global AIDS conference held in Bangkok reveal that the percentage of HIV-infected women is rising on every continent and in every major region in the world except Western Europe and Australia. Young women comprise 75 percent of those between the ages of 15 to 24 who are infected with HIV globally.

*(Continued on next page...)*

### Factors Contributing to Equality

A broader reform agenda that includes **good governance** and **low unemployment** levels is essential to raising the status of women in many countries. International development experts emphasize that while good governance need not fit a Western democratic mold, it must deliver stability through inclusiveness and accountability. Reducing unemployment levels is crucial because countries already unable to provide employment for male job-seekers are not likely to improve employment opportunities for women.

**The spread of information and communication technologies** (ICT) offers great promise. According to World Bank analysis, increases in the level of ICT infrastructure tend to improve gender equality in education and employment. ICT also will enable women to form social and political networks. For regions suffering political oppression, particularly in the Middle East, these networks could become a 21$^{st}$ century counterpart to the 1980s' Solidarity Movement against the Communist regime in Poland.

Women in developing regions often turn to **nongovernmental organizations** (NGOs) to provide basic services. NGOs could become even more important to the status of women by 2020 as women in developing countries face increased threats and acquire IT networking capabilities.

The current trend toward decentralization and devolution of power in most states will afford women **increased opportunities for political participation**. Despite only modest gains in the number of female officeholders at the national level—women currently are heads of state in only eight countries—female participation in local and provincial politics is steadily rising and will especially benefit rural women removed from the political center of a country.

### Other Benefits

The stakes for achieving gender parity are high and not just for women. A growing body of empirical literature suggests that gender equality in education promotes economic growth and reduces child mortality and malnutrition. At the Millennium Summit, UN leaders pledged to achieve gender equity in primary and secondary education by the year 2005 in every country of the world.

- By 2005, the 45 countries that are not on course to meet the UN targets are likely to suffer 1 to 3 percent lower GDP per capita growth as a result.

## Fictional Scenario: Davos World

*This scenario provides an illustration of how robust economic growth over the next 15 years could reshape the globalization process—giving it a more non-Western face. It is depicted in the form of a hypothetical letter from the head of the World Economic Forum to a former US Federal Reserve chairman on the eve of the annual Davos meeting in 2020. Under this scenario, the Asian giants as well as other developing states continue to outpace most "Western" economies, and their huge, consumer-driven domestic markets become a major focus for global business and technology. Many boats are lifted, but some founder. Africa does better than one might think, while some medium-sized emerging countries are squeezed. Western powers, including the United States, have to contend with job insecurity despite the many benefits to be derived from an expanding global economy. Although benefiting from energy price increases, the Middle East lags behind and threatens the future of globalization. In addition, growing tensions over Taiwan may be on the verge of triggering an economic meltdown. At the end of the scenario, we identify some lessons to be drawn from our fictional account, including the need for more management by leaders lest globalization slip off the rails.*

January 12, 2020

Dear Mr. Chairman:

As you know, the last few years have been rough. I finally persuaded the Asians to drop their boycott, and this year we're meeting in China instead of Davos. From now on it will be Switzerland every other year and Asia in the alternate years. I thought at first that I could get the Asians to back down, but they are united. Even the Japanese were not willing to bend. I'm not convinced this was all one big Chinese plot as some are charging. I'm not even sure whether the Chinese were fully in favor of it. Once it caught hold, they had to show some leadership and support Asian claims, but I think they are so confident of their current status that meeting every year in Davos did not bother them. Hosting the sessions actually puts pressure on them to make concessions and deal with some of the complaints about how they do business.

This reminds me of a particular theme I've been developing in my mind as I reflect on how globalization has now evolved. At the turn of the century, we equated globalization with Americanization. America was the model. Now globalization has more of an Asian face and, to be frank, America is no longer quite the engine it used to be. Instead the markets are now oriented eastwards.

That's not to say that the system runs on its own. Only after learning a couple of tough lessons did we see how much management was involved or how easily globalization could come off the rails. We business leaders have had to learn to step in more aggressively.

The 9/11 tragedy was a wake-up call. Terrorism still poses a physical and strategic challenge. In order to protect ourselves, we had to put up barriers, but there was a danger that we would do so much that we would undermine the very basis of globalization—the free flow of capital, goods, people, etc. We tried to strike a delicate balance between security and openness. There's been a lot of criticism about US visa restrictions cutting back on the number of foreign students, and American scientists worried about the US's science and technology leadership slipping away to Asia.

This gets me to my second point. Ten or 15 years ago we did not realize the extent to which the Asian giants were ready to take up the slack. The Chinese and Indians have really maintained the momentum behind globalization. It started out as a US-China dynamic, but now the Asian market is self-generating and not so dependent on trade with the US. Moreover, the competition between China and India over energy supplies and markets has spurred further growth and innovation.

But we had a few sleepless nights over the years, particularly when China ran into financial problems. The fact that the recovery was quick was probably crucial. I think Beijing would have had trouble coping with a full-blown political crisis. Such turmoil could have stymied its economic rise for a decade or more. Fortunately that did not happen. Although the US helped, the really interesting thing was that China dug itself out without the kind of US or international help we thought it would need. Again we underestimated the extent to which China had created a domestic market that could jumpstart its economy.

What the downturn unfortunately did was ignite the latent nationalism that had been lurking below the surface, again increasing tensions over Taiwan. China has been "feeling its oats" and the risk of miscalculation is growing. I'm getting more and more worried as no one—government or private sector—is stepping into the breach to head off what could be a major security and business crisis.

Tensions were also on the rise between China and India and the other emerging states. The success of the Asian giants made it harder for the smaller guys to catch up. The huge pull from China and India on jobs was not just felt in the West. Now we see higher pay for China's workers finally leading to jobs being exported again to lower-wage economies. In part, this can be attributed to demographics—China is a country that is suddenly looking older, its one-child policy coming back to haunt it.

Early on, the outcry in the West over outsourcing and migration could have stalled globalization, but what can we really do—hold back the "tides" of progress in some rerun of Luddite madness? I detected below the surface a strong temptation in Washington and European capitals to play off the emerging countries against China and India by giving preference to non-Chinese products.

On the positive side, it was high-tech breakthroughs that put some countries on the road to sustainable economic growth. Expanded food production from biotechnology innovations and clean water from better filtration systems were boons that helped eliminate the direst poverty and start an export-driven agricultural sector. China and the US finally ganged up on Europe about GMOs.

Higher commodity prices also have been a godsend—much more so than any debt forgiveness scheme. A couple of the Asian-backed energy consortiums practically run two or three of the smaller states. They're popular because they provide not only their workers but all the surrounding communities with full heath-care. Malaria and TB—not to mention AIDS—are being tackled. I'm reminded that businesses—if one thinks back to the East India Company's total rule over the subcontinent in the eighteenth century —were at the forefront when globalization first got going. Have we come full circle with business taking over again from government?

We've seen some progress in the Middle East with a couple countries actually undertaking market liberalization reforms, but others are still stuck in a rut. Palestine yearns for a George Soros figure who can inject a lot of capital and develop an export outlet, but I don't see anyone willing to make the investment.

Elsewhere, revenues generated by high oil prices have enabled the Saudis and others to stem what has been a plunging standard of living for most of them. That's not good in the long run. I fear there's more to this story that we may not like.

Davos has done a lot, I think, in opening up the old exclusive Western club. I admit at first I did not really see it coming—the fact that China and India with their burgeoning middle classes had begun to create such large markets. In the last few years, the whole balance—as I now realize—has been shifting. Asian consumers are setting the trends, and Western businesses have to respond if they want to grow. Fifteen years ago, few of us knew anything about Asian firms. Now we have Wumart. China also got Washington's attention when it started diversifying its foreign currency holdings and the US public awakened to the fact that it had been living way beyond its means.

By itself, Europe probably would have felt threatened by Asia's rapid rise, but—funny thing—a rising Asia was seen as a counterbalance to a dominant US. Asia's growth also helped Europe get out of its slump. The EU thinks it and China have a lot in common—reverence for regional institutions. China with its Shanghai Cooperation Organization, for example. I'm not so sure.

By the way, I heard that your granddaughter is also spending a semester in China, learning the language. Did you know that one of my grandsons is also there? Perhaps we can get the two of them together at the Davos-in-China meeting.

### "Lessons Learned"

This scenario illustrates the vast changes that would be likely to result from continued robust economic growth and the stresses and strains that could derail it.

- Growth in Asian markets would force domestic adjustments on the US and other Western countries that would need to be managed.

- If the global trading system became more integrated and complicated, it would be important to bring China, India and other emerging states more inside the tent, but this would require patience and potential trade-offs.

- It is unlikely that the system would be self-regulating. A strong global economy, for example, would not lead automatically to a resolution of crises like Taiwan.

# Rising Powers: The Changing Geopolitical Landscape

The likely emergence of China and India as new major global players—similar to the rise of Germany in the 19th century and the United States in the early 20th century—will transform the geopolitical landscape, with impacts potentially as dramatic as those of the previous two centuries. In the same way that commentators refer to the 1900s as the "American Century," the early 21st century may be seen as the time when some in the developing world, led by China and India, come into their own.

- The population of the region that served as the locus for most 20th-century history—Europe and Russia—will decline dramatically in relative terms; almost all population growth will occur in developing nations that until recently have occupied places on the fringes of the global economy (see graphic on page 48).[6]

- The "arriviste" powers—China, India, and perhaps others such as Brazil and Indonesia—could usher in a new set of international alignments, potentially marking a definitive break with some of the post-World War II institutions and practices.

- Only an abrupt reversal of the process of globalization or a major upheaval in these countries would prevent their rise. Yet how China and India exercise their growing power and whether they relate cooperatively or competitively to other powers in the international system are key uncertainties.

A combination of sustained high economic growth, expanding military capabilities, active promotion of high technologies, and large populations will be at the root of the expected rapid rise in economic and political power for both countries.

- Because of the sheer size of China's and India's populations—projected by the US Census Bureau to be 1.4 billion and almost 1.3 billion respectively by 2020—their standard of living need not approach Western levels for these countries to become important economic powers.

- China, for example, is now the third largest producer of manufactured goods, its share having risen from four to 12 percent in the past decade. It should easily surpass Japan in a few years, not only in share of manufacturing but also of the world's exports. Competition from "the China price" already powerfully restrains manufactures prices worldwide.

India currently lags behind China (see box on page 53) on most economic measures, but most economists believe it also will sustain high levels of economic growth.

---

[6] CIA, *Long-Term Global Demographic Trends: Reshaping the Geopolitical Landscape*, July 2001.

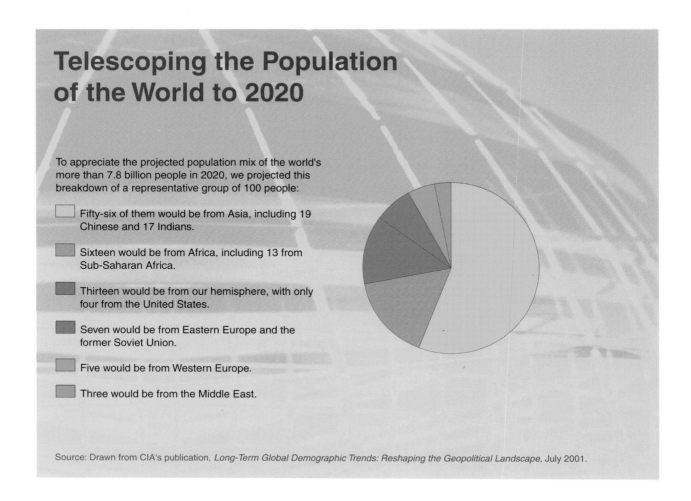

# Telescoping the Population of the World to 2020

To appreciate the projected population mix of the world's more than 7.8 billion people in 2020, we projected this breakdown of a representative group of 100 people:

- Fifty-six of them would be from Asia, including 19 Chinese and 17 Indians.

- Sixteen would be from Africa, including 13 from Sub-Saharan Africa.

- Thirteen would be from our hemisphere, with only four from the United States.

- Seven would be from Eastern Europe and the former Soviet Union.

- Five would be from Western Europe.

- Three would be from the Middle East.

Source: Drawn from CIA's publication, *Long-Term Global Demographic Trends: Reshaping the Geopolitical Landscape,* July 2001.

At the same time, other changes are likely to shape the new landscape. These include the possible economic rise of other states—such as Brazil, South Africa, Indonesia, and even Russia—which may reinforce the growing role of China and India even though by themselves these other countries would have more limited geopolitical impact. Finally, we do not discount the possibility of a stronger, more united Europe and a more internationally activist Japan, although Europe, Japan, and Russia will be hard pressed to deal with aging populations.

The growing demand for energy will drive many of these likely changes on the geopolitical landscape. China's and India's perceived need to secure access to energy supplies will propel these countries to become more global rather than just regional powers, while Europe and Russia's co-dependency is likely to be strengthened.

## Rising Asia

***China's*** desire to gain "great power" status on the world stage will be reflected in its greater economic leverage over countries in the region and elsewhere as well as its steps to strengthen its military. East Asian states are adapting to the advent of a more powerful China by forging closer economic and political ties with Beijing, potentially accommodating themselves to its preferences, particularly on sensitive issues like Taiwan.

- Japan, Taiwan, and various Southeast Asian nations, however, also may try to appeal to each other and the United States to counterbalance China's growing influence.

China will continue to strengthen its military through developing and acquiring modern weapons, including advanced fighter aircraft, sophisticated submarines, and increasing numbers of ballistic missiles. China will overtake Russia and others as the second largest defense spender after the United States over the next two decades and will be, by any measure, a first-rate military power.

Economic setbacks and crises of confidence could slow China's emergence as a full-scale great power, however. Beijing's failure to maintain its economic growth would itself have a global impact.

- Chinese Government failure to satisfy popular needs for job creation could trigger political unrest.

- Faced with a rapidly aging society beginning in the 2020s, China may be hard pressed to deal with all the issues linked to such serious demographic problems. It is unlikely to have developed by then the same coping mechanisms—such as sophisticated pension and health-care systems—characteristic of Western societies.

- If China's economy takes a downward turn, regional security would weaken, resulting in heightened prospects for political instability, crime, narcotics trafficking, and illegal migration.

*"Economic setbacks and crises of confidence could slow China's emergence as a full-scale great power…."*

The rise of **India** also will present strategic complications for the region. Like China, India will be an economic magnet for the region, and its rise will have an impact not only in Asia but also to the north—Central Asia, Iran, and other countries of the Middle East. India seeks to bolster regional cooperation both for strategic reasons and because of its desire to increase its leverage with the West, including in such organizations as the World Trade Organization (WTO).

# China's Rise

## Stock of Inward Foreign Direct Investment, 1980-2003

*China has risen from negligible in 1980 to a 6-percent share. . . could be in second place by 2020.*

Trillion US $

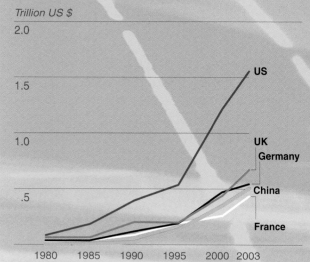

Source: UN, *World Investment Report 2004*.

## Share of World Manufacturing, 1980-2003

*It could easily surpass Japan in a few years.*

Percent

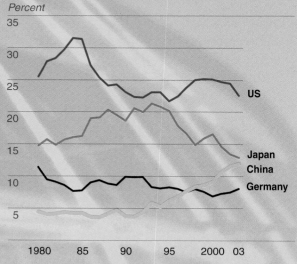

Source: Global Insight.

## Share of World's Exports, 1980-2003

*It should also overtake Japan in trade . . .*

Percent

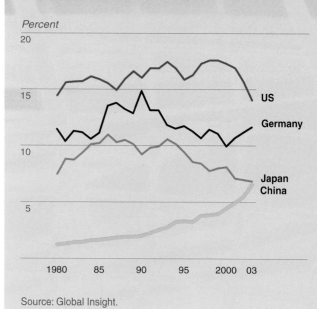

Source: Global Insight.

## Growth in Demand for Oil, 2000-2020

*. . . and nearly match the US as a dominant driver of additional oil demand.*

Million barrels/day

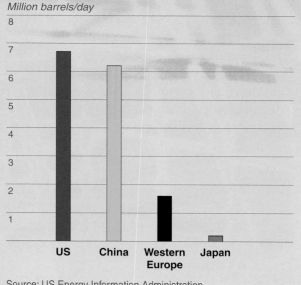

Source: US Energy Information Administration.

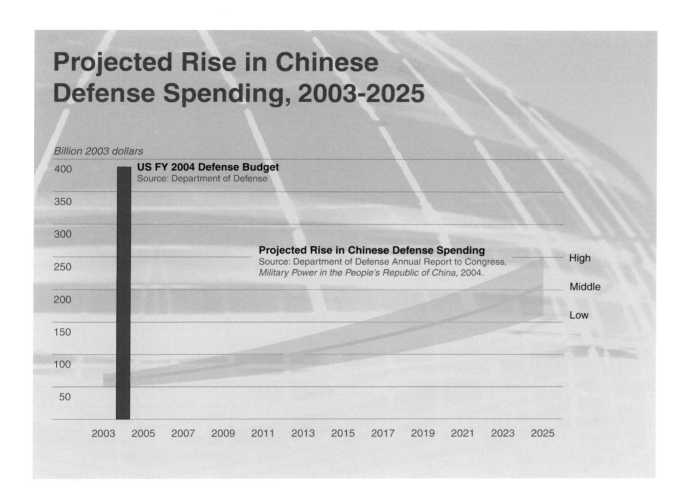

**Projected Rise in Chinese Defense Spending, 2003-2025**

*Billion 2003 dollars*

**US FY 2004 Defense Budget**
Source: Department of Defense

**Projected Rise in Chinese Defense Spending**
Source: Department of Defense Annual Report to Congress, *Military Power in the People's Republic of China, 2004.*

High

Middle

Low

As India's economy grows, governments in Southeast Asia—Malaysia, Singapore, Thailand, and other countries—may move closer to India to help build a potential geopolitical counterweight to China. At the same time, India will seek to strengthen its ties with countries in the region without excluding China.

- Chinese-Indian bilateral trade is expected to rise rapidly from its current small base of $7.6 billion, according to Goldman Sachs and other experts.

Just like China, India may stumble and experience political and economic volatility with pressure on resources—

land, water, and energy supplies—intensifying as it modernizes. For example, India will face stark choices as its population increases and its surface and ground water become even more polluted.

**Other Rising States?**

Brazil, Indonesia, Russia, and South Africa also are poised to achieve economic growth, although they are unlikely to exercise the same political clout as China or India. Their growth undoubtedly will benefit their neighbors, but they appear unlikely to become such economic engines that they will be able to alter the flow of economic power within and through their

**Risks to Chinese Economic Growth**

Whether China's rise occurs smoothly is a key uncertainty.  In 2003, the RAND Corporation identified and assessed eight major risks to the continued rapid growth of China's economy over the next decade.  Its "Fault Lines in China's Economic Terrain" highlighted:

- Fragility of the financial system and state-owned enterprises

- Economic effects of corruption

- Water resources and pollution

- Possible shrinkage of foreign direct investment

- HIV/AIDS and epidemic diseases

- Unemployment, poverty, and social unrest

- Energy consumption and prices

- Taiwan and other potential conflicts

RAND's estimates of the negative growth impact of these adverse developments occurring separately on a one-at-a-time basis range from a low of between 0.3 and 0.8 percentage points for the effects of poverty, social unrest, and unemployment to a high of between 1.8 and 2.2 percentage points for epidemic disease.

- The study assessed the probability that *none* of these developments would occur before 2015 as low and noted that they would be more likely to occur in clusters rather than individually – financial distress, for example, would also worsen corruption, compound unemployment, poverty, and social unrest, and reduce foreign direct investment.

- RAND assessed the probability of *all* of these adverse developments occurring before 2015 as very low but estimated that should they all occur their cumulative effect would be to reduce Chinese economic growth by between 7.4 and 10.7 percentage points—effectively wiping out growth during that time frame.

**India vs. China:  Long-Term Prospects**

India lags economically behind China, according to most measures such as overall GDP, amount of foreign investment (today a small fraction of China's), and per capita income.  In recent years, India's growth rate has lagged China's by about 20 percent.  Nevertheless, some experts believe that India might overtake China as the fastest growing economy in the world.  India has several factors working for it:

- Its working-age population will continue to increase well into the 2020s, whereas, due to the one-child policy, China's will diminish and age quite rapidly.

- India has well-entrenched democratic institutions, making it somewhat less vulnerable to political instability, whereas China faces the continuous challenge of reconciling an increasingly urban and middle-class population with an essentially authoritarian political system.

- India possesses working capital markets and world-class firms in some important high-tech sectors, which China has yet to achieve.

On the other hand, while India has clearly evolved beyond what the Indians themselves referred to as the 2-3 percent "Hindu growth rate," the legacy of a stifling bureaucracy still remains.  The country is not yet attractive for foreign investment and faces strong political challenges as it continues down the path of economic reform.  India is also faced with the burden of having a much larger proportion of its population in desperate poverty.  In addition, some observers see communal tensions just below the surface, citing the overall decline of secularism, growth of regional and caste-based political parties, and the 2002 "pogrom" against the Muslim minority in Gujarat as evidence of a worsening trend.

Several factors could weaken China's prospects for economic growth, especially the risks to political stability and the challenges facing China's financial sector as it moves toward a fuller market orientation.  China might find its own path toward an "Asian democracy" that may not involve major instability or disruption to its economic growth—but there are a large number of unknowns.

In many other respects, both China and India still resemble other developing states in the problems each must overcome, including the large numbers, particularly in rural areas, who have not enjoyed major benefits from economic growth.  Both also face a potentially serious HIV/AIDS epidemic that could seriously affect economic prospects if not brought under control.  According to recent UN data, India has overtaken South Africa as the country with the largest number of HIV-infected people.

The bottom line:  India would be hard-pressed to accelerate economic growth rates to levels above those reached by China in the past decade.  But China's ability to sustain its current pace is probably more at risk than is India's; should China's growth slow by several percentage points, India could emerge as the world's fastest-growing economy as we head towards 2020.

regions—a key element in Beijing and New Delhi's political and economic rise.

Experts acknowledge that **Brazil** is a pivotal state with a vibrant democracy, a diversified economy and an entrepreneurial population, a large national patrimony, and solid economic institutions. Brazil's success or failure in balancing pro-growth economic measures with an ambitious social agenda that reduces poverty and income inequality will have a profound impact on region-wide economic performance and governance during the next 15 years. Luring foreign direct investment and advancing regional stability and equitable integration—including trade and economic infrastructure—probably will remain axioms of Brazilian foreign policy. Brazil is a natural partner both for the United States and Europe and for rising powers China and India and has the potential to enhance its leverage as a net exporter of oil.

Experts assess that over the course of the next decade and a half **Indonesia** may revert to high growth of 6 to 7 percent, which along with its expected increase in its relatively large population from 226 to around 250 million would make it one of the largest developing economies. Such high growth would presume an improved investment environment, including intellectual property rights protection and openness to foreign investment. With slower growth its economy would be unable to absorb the unemployed or under-employed labor force, thus heightening the risk of greater

political instability. Indonesia is an amalgam of divergent ethnic and religious groups. Although an Indonesian national identity has been forged in the five decades since independence, the government is still beset by stubborn secessionist movements.

**Russia's** energy resources will give a boost to economic growth, but Russia faces a severe demographic challenge resulting from low birth rates, poor medical care, and a potentially explosive AIDS situation. US Census Bureau projections show the working-age population likely to shrink dramatically by 2020. Russia's present trajectory away from pluralism toward bureaucratic authoritarianism also decreases the chances it will be able to attract foreign investment outside the energy sector, limiting prospects for diversifying its economy. The problems along its southern borders—including Islamic extremism, terrorism, weak states with poor governance, and conflict—are likely to get worse over the next 15 years. Inside Russia, the autonomous republics in North Caucasus risk failure and will remain a source of endemic tension and conflict. While these social and political factors limit the extent to which Russia can be a major global player, in the complex world of 2020 Russia could be an important, if troubled, partner both for the established powers, such as the United States and Europe, and the rising powers of China and India. The potential also exists for Russia to enhance its leverage with others as a result of its position as a major oil and gas exporter.

## Asia: The Cockpit for Global Change?

According to the regional experts we consulted, Asia will exemplify most of the trends that we see as shaping the world over the next 15 years. Northeast and Southeast Asia will progress along divergent paths—the countries of the North will become wealthier and more powerful, while at least some states in the South may lag economically and will continue to face deep ethnic and religious cleavages. As Northeast Asia acts as a political and economic center of gravity for the countries of the South, parts of Southeast Asia will be a source of transnational threats—terrorism and organized crime—to the countries of the North. The North/South divisions are likely to be reflected in a cultural split between non-Muslim Northeast Asia, which will adapt to the continuing spread of globalization, and Southeast Asia, where Islamic fundamentalism may increasingly make inroads in such states as Indonesia, Malaysia, and parts of The Philippines. The diversion of investment towards China and India also could spur Southeast Asia to implement plans for a single economic community and investment area by 2020.

The experts also felt that demographic factors will play a key role in shaping regional developments. China and other countries in Northeast Asia, including South Korea, will experience a slowing of population growth and a "graying" of their peoples over the next 15 years. China also will have to face the consequences of a gender imbalance caused by its one-child policy. In Southeast Asian countries such as The Philippines and Indonesia, rising populations will challenge the capacity of governments to provide basic services. Population and poverty pressures will spur migration within the region and to Northeast Asia. High population concentrations and increasing ease of travel will facilitate the spread of infectious diseases, risking the outbreak of pandemics.

The regional experts felt that the possibility of major inter-state conflict remains higher in Asia than in other regions. In their view, the Korean Peninsula and Taiwan Strait crises are likely to come to a head by 2020, risking conflict with global repercussions. At the same time, violence within Southeast Asian states—in the form of separatist insurgencies and terrorism—could intensify. China also could face sustained armed unrest from separatist movements along its western borders.

Finally, the roles of and interaction between the region's major powers—China, Japan, and the US—will undergo significant change by 2020. The United States and China have strong incentives to avoid confrontation, but rising nationalism in China and fears in the US of China as an emerging strategic competitor could fuel an increasingly antagonistic relationship. Japan's relationship with the US and China will be shaped by China's rise and the nature of any settlement on the Korean Peninsula and over Taiwan.

*"Russia's energy resources will give a boost to economic growth, but Russia faces a severe demographic challenge…[with its] working-age population likely to shrink dramatically."*

**South Africa** will continue to be challenged by AIDS and widespread crime and poverty, but prospects for its economy—the largest in the region—look promising. According to some forecasts, South Africa's economy is projected to grow over the next decade in the 4- to 5-percent range if reformist policies are implemented. Experts disagree over whether South Africa can be an engine for more than southern Africa or will instead forge closer relationships with middling or up-and-coming powers on other continents. South African experts adept at scenario-building and gaming see the country's future as lying with partnerships formed outside the region.

## The "Aging" Powers

**Japan**'s economic interests in Asia have shifted from Southeast Asia toward Northeast Asia—especially China and the China-Japan-Korea triangle—over the past two decades and experts believe the aging of Japan's work force will reinforce dependence on outbound investment and greater economic integration with Northeast Asia, especially China[7]. At the same time, Japanese concerns regarding regional stability are likely to grow owing to the ongoing crisis over North Korea, continuing tensions between China and Taiwan and the challenge of integrating rising China and India without major disruption. If anything, growing Chinese economic power is likely to spur increased activism by Japan on the world stage.

Opinion polls indicate growing public support for Japan becoming a more "normal" country with a proactive foreign policy. Experts see various trajectories that Japan could follow depending on such factors as the extent of China's growing strength, a resurgence or lack of continued vitality in Japan's economy, the level of US influence in the region and how developments in Korea and Taiwan play out. At some point, for example, Japan may have to choose between "balancing" against or "bandwagoning" with China.

*"…Europe's strength may be in providing… a model of global and regional governance to the rising powers…"*

By most measures—market size, single currency, highly skilled work force, stable democratic governments, unified trade bloc, and GDP—**an enlarged Europe** will have the ability to increase its weight on the international scene. Its crossroads location and the growing diversity of its population—particularly in pulling in new members—provides it with a unique ability to forge strong bonds both to the south with the Muslim world and Africa and to the east with Russia and Eurasia.

---

[7] *Asia's Shifting Strategic Landscape*, Foreign Policy Research Institute, 26 November 2003.

The extent to which Europe enhances its clout on the world stage depends on its ability to achieve greater political cohesion. In the short term, taking in ten new east European members probably will be a "drag" on the deepening of European Union (EU) institutions necessary for the development of a cohesive and shared "strategic vision" for the EU's foreign and security policy.

- Unlike the expansion when Ireland, Spain, Portugal and Greece joined the Common Market in the 1970s and early 1980s, Brussels has a fraction of the structural funds available for quickly bringing up the Central Europeans to the economic levels of the rest of the EU.

- Possible Turkish membership presents both challenges—because of Turkey's size and religious and cultural differences—as well as opportunities, provided that mutual acceptance and agreement can be achieved. In working through the problems, a path might be found that can help Europe to accommodate and integrate its growing Muslim population.

Defense spending by individual European countries, including the UK, France, and Germany is likely to fall further behind China and other countries over the next 15 years. Collectively these countries will outspend all others except the US and possibly China[8]. EU member states historically have had difficulties in coordinating and rationalizing defense spending in such a way as to boost capabilities despite progress on a greater EU security and defense role. Whether the EU will develop an army is an open question, in part because its creation could duplicate or displace NATO forces.

While its military forces have little capacity for power projection, Europe's strength may be in providing, through its commitment to multilateralism, a model of global and regional governance to the rising powers, particularly if they are searching for a "Western" alternative to strong reliance on the United States. For example, an EU-China alliance, though still unlikely, is no longer unthinkable.

Aging populations and shrinking work forces in most countries will have an important impact on the continent, creating a serious but not insurmountable economic and political challenge. Europe's total fertility rate is about 1.4— well below the 2.1 replacement level. Over the next 15 years, West European economies will need to find several million workers to fill positions vacated by retiring workers. Either European countries adapt their work forces, reform their social welfare, education, and tax systems, and accommodate growing immigrant populations (chiefly from Muslim countries) or they face a period of protracted economic stasis that could threaten the huge successes made in creating a more United Europe.

---

[8] *Strategic Trends*, Joint Doctrine and Concepts Centre, March 2003.

## Global Aging and Migration

According to US Census Bureau projections, about half of the world's population lives in countries or territories whose fertility rates are not sufficient to replace their current populations. This includes not only Europe, Russia, and Japan, where the problem is particularly severe, but also most parts of developed regions such as Australia, New Zealand, North America, and East Asian countries like Singapore, Hong Kong, Taiwan, and South Korea. Certain countries in the developing world, including Arab states such as Turkey, Algeria, Tunisia, and Lebanon, also are dropping below the level of 2.1 children per woman necessary to maintain long-term population stability.[9]

China is a special case where the transition to an aging population—nearly 400 million Chinese will be over 65 by 2020—is particularly abrupt and the emergence of a serious gender imbalance could have increasing political, social, and even international repercussions. An unfunded nationwide pension arrangement means many Chinese may have to continue to work into old age.

Migration has the potential to help solve the problem of a declining work force in Europe and, to a lesser degree, Russia and Japan and probably will become a more important feature of the world of 2020, even if many of the migrants do not have legal status. Recipient countries face the challenge of integrating new immigrants so as to minimize potential social conflict.

- Remittances from migrant workers are increasingly important to developing economies. Some economists believe remittances are greater than foreign direct investment in most poor countries and in some cases are more valuable than exports.

However, today one-half of Nigerian-born medical doctors and PhDs reside in the United States. Most experts do not expect the current, pronounced trend of "brain drain" from the Middle East and Africa to diminish. Indeed, it could increase with the expected growth of employment opportunities, particularly in Europe.

---

[9] Nicholas Eberstadt, "Four Surprises in Global Demography," Foreign Policy Research Institute's *Watch on the West*, Vol 5, Number 5, July 2004.

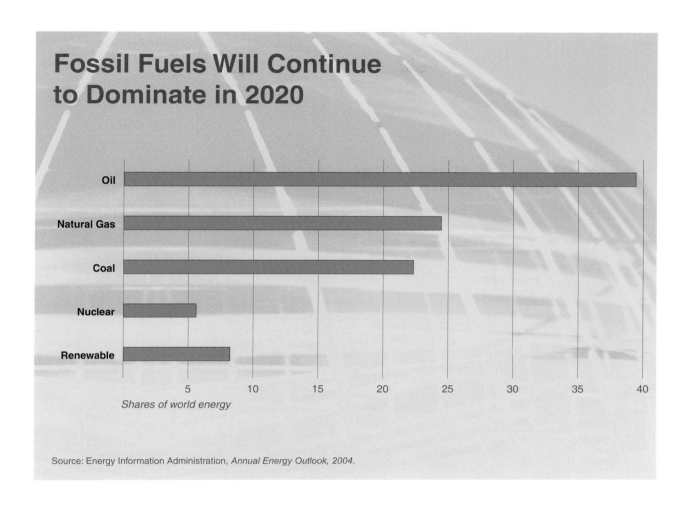

**Fossil Fuels Will Continue to Dominate in 2020**

*Shares of world energy*

Source: Energy Information Administration, *Annual Energy Outlook, 2004*.

## Growing Demands for Energy

Growing demands for energy—especially by the rising powers—through 2020 will have substantial impacts on geopolitical relations. The single most important factor affecting the demand for energy will be global economic growth, particularly that of China and India.

- Despite the trend toward more efficient energy use, total energy consumed probably will rise by about 50 percent in the next two decades compared to a 34 percent expansion from 1980–2000, with an increasing share provided by petroleum.

- Renewable energy sources such as hydrogen, solar, and wind energy probably will account for only about 8 percent of the energy supply in 2020. While Russia, China, and India all plan expansions of their nuclear power sector, nuclear power probably will decline globally in absolute terms in the next decade.

The International Energy Agency assesses that with substantial investment in new capacity, overall energy supplies will be sufficient to meet growing global demand. Continued

# An Expanding European Union

| Current | European Union (EU) 15 | EU 25 | EU 30 | China | Russia | US |
|---|---|---|---|---|---|---|
| **Population** (*million*) | 380 | 458 | 527 | 1,299 | 144 | 293 |
| **GDP** (*billion USD*) | 11,000 | 11,637 | 12,091 | 6,449 | 1,287 | 10,980 |
| **Area** (*1,000 mi$^2$*) | 1,212 | 1,493 | 1,789 | 3,646 | 6,488 | 3,659 |

Source: CIA.

## Could Europe Become A Superpower?

According to the regional experts we consulted, Europe's future international role depends greatly on whether it undertakes major structural economic and social reforms to deal with its aging work-force problem. The demographic picture will require a concerted, multidimensional approach including:

- ***More legal immigration and better integration of workers likely to be coming mainly from North Africa and the Middle East***. Even if more guest workers are not allowed in, Western Europe will have to integrate a growing Muslim population. Barring increased legal entry may only lead to more illegal migrants who will be harder to integrate, posing a long-term problem. It is possible to imagine European nations successfully adapting their work forces and social welfare systems to these new realities; it is harder to see a country—Germany, for example—successfully assimilating millions of new Muslim migrant workers in a short period of time.

- ***Increased flexibility in the workplace, such as encouraging young women to take a few years off to start families in return for guarantees of reentry***. Encouraging the "younger elderly" (50-65 year olds) to work longer or return to the work force also would help ease labor shortages.

The experts felt that the current welfare state is unsustainable and the lack of any economic revitalization could lead to the splintering or, at worst, disintegration of the European Union, undermining its ambitions to play a heavyweight international role.

The experts believe that the EU's economic growth rate is dragged down by Germany and its restrictive labor laws. Structural reforms there—and in France and Italy to lesser extents—remain key to whether the EU as a whole can break out of its slow-growth pattern. A total break from the post-World War II welfare state model may not be necessary, as shown in Sweden's successful example of providing more flexibility for businesses while conserving many worker rights. Experts are dubious that the present political leadership is prepared to make even this partial break, believing a looming budgetary crisis in the next five years would be the more likely trigger for reform.

If no changes were implemented Europe could experience a further overall slowdown, and individual countries might go their own way, particularly on foreign policy, even if they remained nominal members. In such a scenario, enlargement is likely to stop with current members, making accession unlikely for Turkey and the Balkan countries, not to mention long-term possibilities such as Russia or Ukraine. Doing just enough to keep growth rates at one or two percent may result in some expansion, but Europe probably would not be able to play a major international role commensurate with its size.

In addition to the need for increased economic growth and social and welfare reform, many experts believe the EU has to continue streamlining the complicated decision-making process that hinders collective action. A *federal* Europe—unlikely in the 2020 timeframe—is not necessary to enable it to play a weightier international role so long as it can begin to mobilize resources and fuse divergent views into collective policy goals. Experts believe an economic "leap forward"—stirring renewed confidence and enthusiasm in the European project—could trigger such enhanced international action.

limited access of the international oil companies to major fields could restrain this investment, however, and many of the areas—the Caspian Sea, Venezuela, West Africa and South China Sea—that are being counted on to provide increased output involve substantial political or economic risk. Traditional suppliers in the Middle East are also increasingly unstable. Thus sharper demand-driven competition for resources, perhaps accompanied by a major disruption of oil supplies, is among the key uncertainties.

*China* and *India*, which lack adequate domestic energy resources, will have to ensure continued access to outside suppliers; thus, the need for energy will be a major factor in shaping their foreign and defense policies, including expanding naval power.

- Experts believe China will need to boost its energy consumption by about 150 percent and India will need to nearly double its consumption by 2020 to maintain a steady rate of economic growth.

- Beijing's growing energy requirements are likely to prompt China to increase its activist role in the world—in the Middle East, Africa, Latin America, and Eurasia. In trying to maximize and diversify its energy supplies, China worries about being vulnerable to pressure from the United States which Chinese officials see as having an aggressive energy policy that can be used against Beijing.

- For more than ten years Chinese officials have openly asserted that production from Chinese firms

*The Geopolitics of Gas.* Both oil and gas suppliers will have greater leverage than today, but the relationship between gas suppliers and consumers is likely to be particularly strong because of the restrictions on delivery mechanisms. Gas, unlike oil, is not yet a fungible source of energy, and the interdependency of pipeline delivery—producers must be connected to consumers, and typically neither group has many alternatives—reinforces regional alliances.

- More than 95 percent of gas produced and three quarters of gas traded is distributed via pipelines directly from supplier to consumer, and gas-to-liquids technology is unlikely to change these ratios substantially by 2020.

- Europe will have access to supplies in Russia and North Africa while China will be able to draw from eastern Russia, Indonesia, and potentially huge deposits in Australia. The United States will look almost exclusively to Canada and other western hemisphere suppliers.

investing overseas is more secure than imports purchased on the international market. Chinese firms are being directed to invest in projects in the Caspian region, Russia, the Middle East, and South America in order to secure more reliable access.

*Europe's* energy needs are unlikely to grow to the same extent as those of the developing world, in part because of Europe's expected lower economic growth and more efficient use of energy.

Europe's increasing preference for natural gas, combined with depleting reserves in the North Sea, will give an added boost to political efforts already under way to strengthen ties with Russia and North Africa, as gas requires a higher level of political commitment by both sides in designing and constructing the necessary infrastructure. According to a study by the European Commission, the Union's share of energy from foreign sources will rise from about half in 2000 to two-thirds by 2020. Gas use will increase most rapidly due to environmental concerns and the phasing out of much of the EU's nuclear energy capacity.

*"...many of the areas... being counted on to provide increased [energy] output involve substantial political or economic risk.... Thus sharper demand-driven competition... perhaps accompanied by a major disruption of oil supplies, is among the key uncertainties."*

Deliveries from the Yamal-Europe pipeline and the Blue Stream pipeline will help **Russia** increase its gas sales to the EU and Turkey by more than 40 percent over 2000 levels in the first decade of the 21$^{st}$ century; as a result, Russia's share of total European demand will rise from 27 percent in 2000 to 31 percent in 2010. Russia, moreover, as the largest energy supplier outside of OPEC, will be well positioned to marshal its oil and gas reserves to support domestic and foreign policy objectives. **Algeria** has the world's eighth largest gas reserves and also is seeking to increase its exports to Europe by 50 percent by the end of the decade.

## US Unipolarity—How Long Can It Last?

A world with a single superpower is unique in modern times. Despite the rise in anti-Americanism, most major powers today believe countermeasures such as balancing are not likely to work in a situation in which the US controls so many of the levers of power. Moreover, US policies are not perceived as sufficiently threatening to warrant such a step.

- Growing numbers of people around the world, especially in the Middle East and the broader Muslim world, believe the US is bent on regional domination—or direct political and economic domination of other states and their resources. In the future, growing distrust could prompt governments to take a more hostile approach, including resistance to support for US interests in multinational forums and development of asymmetric military capabilities as a hedge against the US.

*"There are few policy-relevant theories to indicate how states are likely to deal with a situation in which the US continues to be the single most powerful actor economically, militarily, and technologically."*

Most countries are likely to experiment with a variety of different tactics from various degrees of resistance to engagement in an effort to influence how US power is exercised. We expect that countries will pursue strategies designed to exclude or isolate the US—perhaps temporarily—in order to force or cajole the US into playing by others' rules. Many countries increasingly believe that the surest way to gain leverage over Washington is by threatening to withhold cooperation.  In other forms of bargaining, foreign governments will try to find ways to "bandwagon" or connect their policy agendas to those of the US—for example on the war on terrorism—and thereby fend off US opposition to other policies.

## Fictional Scenario: Pax Americana

**The scenario portrayed below looks at how US predominance may survive radical changes to the global political landscape, with Washington remaining the central pivot for international politics.  It is depicted as the diary entry by a fictitious UN Secretary-General in 2020.  Under this scenario, key alliances and relationships with Europe and Asia undergo change.  US-European cooperation is renewed, including on the Middle East.  There are new security arrangements in Asia, but the United States still does the heavy lifting.  The scenario also suggests that Washington has to struggle to assert leadership in an increasingly diverse, complex, and fast-paced world.  At the end of the scenario, we identify lessons learned from how the scenario played out.**

September 11, 2020

It has been exactly nineteen years today since the view from the 38th floor changed with the destruction of the Twin Towers. I was remarking when the President of the United States phoned that more than the skyline has been altered since. Not only has a new structure been built, partially obscuring the devastation of 9/11, but the US has risen like a phoenix-albeit a beleaguered one- and it again seems to be the bedrock of the international order.

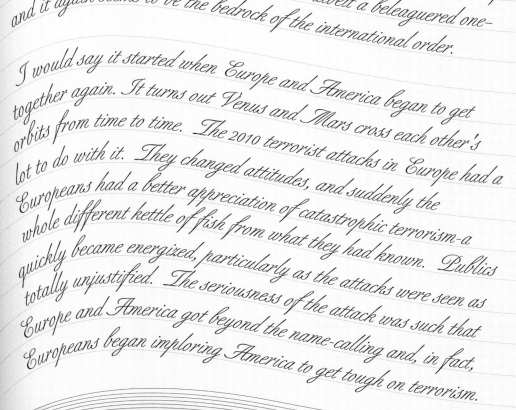

I would say it started when Europe and America began to get together again. It turns out Venus and Mars cross each other's orbits from time to time. The 2010 terrorist attacks in Europe had a lot to do with it. They changed attitudes, and suddenly the Europeans had a better appreciation of catastrophic terrorism-a whole different kettle of fish from what they had known. Publics quickly became energized, particularly as the attacks were seen as totally unjustified. The seriousness of the attack was such that Europe and America got beyond the name-calling and, in fact, Europeans began imploring America to get tough on terrorism.

The closing of transatlantic ranks was prompted by more than this. One thing that became clear is that Europe was more unified than some of our American friends imagined. New Europe turned out to be not that much different from Old Europe, once it joined the EU club and began hanging out in Brussels. NATO went through some rough times but is now working better with the EU. There is grudging acceptance on both sides that NATO has the necessary military tools while the EU can bring to the table a capacity for nation-building.

On the European side, a lot had to do with Turkish accession—something I never expected to see. With the prospect of Turkey coming in, the Europeans realized that their border was now squarely in the Middle East and that meant they had to be more prepared to deal with all the problems of terrorism, fundamentalism, youth bulges, etc.

Coming together as they did, Europe helped to persuade the US that something had to be done to stop the spiraling violence in Palestine. For the Europeans, that had always been the root of the problem, but divisions and a lack of will always got in the way of any concerted action.

Energy and climate change is also playing an increasing role in the US-European dynamics, but not the way one would have expected. For a while, the Europeans looked like they were trying to isolate the US and insist on Washington playing by EU rules. But that was never really going to happen, and European leaders did not factor in their own publics' increasing resentment of China's and other developing countries' flaunting of environmental standards. Kyoto was suddenly out and a new framework had to be thought up with the Americans inside.

The US role changed even more dramatically in Asia. China was rising and, while not directly challenging the US, was certainly displacing it in the region, particularly economically. America's preoccupation with Iraq and terrorism looked set to diminish the US role even more. Japan stood close to the US on Iraq but was conflicted too because of its economic dependence on China. In South Korea, the younger generation blamed the US for the division and problems with the North. It seemed only a matter of time before the US would be pushed to the sidelines.

Then a series of events occurred which changed the dynamic. Frightened by the continuing impasse in North Korea, rumors leaked of the Japanese seriously thinking about their own bomb. About this time China also suffered an economic relapse, which exacerbated the confrontational tone over Taiwan, also heightening worries in Japan and Southeast Asia. The US initially wanted to heighten its profile but found that many worried about a US-China conflict. Washington ended up scaling back its military presence in Korea and Japan. Most do not want the US to leave; even China, I think, secretly sees some virtue in having the US around inasmuch as it makes the others more accepting of Beijing's growing political and economic influence.

This is hardly a marriage made in Heaven and both the US and China have to work hard to keep from going off the rails. I see more storm clouds over Taiwan on the horizon. Nationalism appears to afflict everybody. The rising Chinese middle class turns out to be less interested in democracy and more in nationalism.

Worries about energy supplies have also redounded in America's favor. A stable Middle East is a must for China on energy grounds just as it is for Europe. The US also is increasingly a balancer between the Shia and the Sunnis. Washington may not have been prepared for how a free and Shia-dominated Iraq began to swing the balance and raise tensions right in the region where most of the world's oil comes from. Not an enviable position to be in for the Americans.

I get the feeling at times that a lot of Americans are getting tired of playing the world's policeman. The security burden is still on their shoulders; this is the source of their frustration with the Europeans, who just want to focus on the EU. The Americans thought they had a deal - Washington would do the heavy lifting with Israel, but the Europeans would be prepared to pony up money and troops for a Middle East peacekeeping force. But that looks to be in tatters.

How long will Pax Americana last? I'm not sure. It hasn't meant much institution-building. Sure, the UN works a little better because there is more cooperation among the members, but we've done little on reform. India is getting more frustrated. The Africans, Latin Americans and poor Asians still feel underappreciated and some even are resentful of China and India's rise, which has squeezed out opportunities for them. The Human Rights Commission, which morphed a few years back into the Human Rights and Ethics Commission, is stymied over human cloning, GMOs and whether energy consumption should be globally regulated. For all the practical cooperation there has been on terrorism, we still don't have an agreed-upon definition which could unite countries around a counterterrorism strategy. To top it off, I am in a vicious fight on two fronts in order to stay here in New York, both against the "America firster" groups calling for the UN's removal and large numbers of Europeans and Asians who think the UN is too much under the thumb of Washington.

*I often wonder just how much real progress there has been. I must talk to the US President about this the next time we "girls" get together*

## "Lessons Learned"

A geopolitical environment in which US power dominates would become more complex.

- A rising China might put the US in a different position, having to act as the balancer between China on the one hand and Japan and other Asian countries on the other.

Competing coalitions would be likely to fight over moral and ethical issues. The US would be looked to for leadership, but the scenario suggests it would take dexterity to achieve consensus.

And the Pax Americana would not necessarily be a "sweet" deal for the United States.

- Deeply entrenched expectations about the US providing the heavy lifting on security would likely to be hard to change, particularly given the underlying reality that only the US has the military capacity.

- The scenario suggests that international architecture is not designed for security burdensharing. Other than the US-dominated NATO, no other regional security organizations appear to be operational in the scenario.

# New Challenges to Governance

The nation-state will continue to be the dominant unit of the global order, but economic globalization and the dispersion of technologies, especially information technologies, will place enormous strains on governments. Regimes that were able to manage the challenges of the 1990s could be overwhelmed by those of 2020. Contradictory forces will be at work: authoritarian regimes will face new pressures to democratize, but fragile new democracies may lack the adaptive capacity to survive and develop.

- With migration on the increase in several places around the world—from North Africa and the Middle East into Europe, Latin America and the Caribbean into the United States, and increasingly from Southeast Asia into the northern regions—more countries will be multi-ethnic and multi-religious and will face the challenge of integrating migrants into their societies while respecting their ethnic and religious identities.

**Halting Progress on Democratization**
Global economic growth has the potential to spur democratization, but backsliding by many countries that were considered part of the "third wave" of democratization is a distinct possibility. In particular, by 2020 democratization may be partially reversed among the states of the former Soviet Union and in Southeast Asia, some of which never really embraced democracy. **Russia** and most of the **Central Asian** regimes appear to be slipping back toward authoritarianism, and global economic growth probably will not on its own reverse such a trend. The development of more diversified economies in these countries—by no means inevitable—would be crucial in fostering the growth of a middle class, which in turn would spur democratization.

- Beset already by severe economic inequalities, aging Central Asian rulers must contend with unruly and large youth populations lacking broad economic opportunities. Central Asian governments are likely to suppress dissent and revert to authoritarianism to maintain order, risking growing insurgencies.

*"...backsliding by many countries that were considered part of the 'third wave' of democratization is a distinct possibility."*

**Chinese** leaders will face a dilemma over how much to accommodate pluralistic pressure and relax political controls or risk a popular backlash if they do not. Beijing also has to weigh in the balance its ambition to be a major global player, which would be enhanced if its rulers moved towards political reform.

China may pursue an "Asian way" of democracy that might involve elections at the local level and a consultative mechanism on the national level, perhaps

## Eurasian Countries:  Going Their Separate Ways?

The regional experts who attended our conference felt that Russia's political development since the fall of Communism has been complicated by the continuing search for a post-Soviet national identity.  Putin has increasingly appealed to Russian nationalism—and, occasionally, xenophobia—to define Russian identity.  His successors may well define Russian identity by highlighting Russia's imperial past and its domination over its neighbors even as they reject communist ideology.

In the view of the experts, Central Asian states are weak, with considerable potential for religious and ethnic conflict over the next 15 years.  Religious and ethnic movements could have a destabilizing impact across the region.  Eurasia is likely to become more differentiated despite the fact that demographic counterforces—such as a dearth of manpower in Russia and western Eurasia and an oversupply in Central Asia—could help pull the region together.  Moreover, Russia and the Central Asians are likely to cooperate in developing transportation corridors for energy supplies.

The participants assessed that among the resource-rich countries, Russia has the best prospects for expanding its economy beyond resource extraction and becoming more integrated into the world economy.  To diversify its economy, Russia would need to undertake structural changes and institute the rule of law, which could in turn encourage foreign direct investment outside of the energy sector.  Knowing that Europe probably would want to forge a "special relationship" with a Russia that is stronger economically, Moscow probably would be more tolerant of former Soviet states moving closer to Europe.  If Russia fails to diversify its economy, it could well experience the petro-state phenomenon of unbalanced economic development, huge income inequality, capital flight, and increased social problems.

Regional experts were less confident about the potential for significant economic diversification in the other resource-rich countries in Central Asia and the South Caucasus over the next 15 years—in particular, Kazakhstan, Turkmenistan, and Azerbaijan.  For countries with more limited natural resources, such as Ukraine, Georgia, Kyrgyztan, Tajikistan and Uzbekistan, the challenge will be to develop effective project and service industries, requiring better governance.

Central Asian countries—Kazakhstan, Krgyzstan, Tajikistan, Turkmenistan, and Uzbekistan— face the stiff challenge of keeping the social peace in a context of high population growth, a relatively young population, limited economic prospects, and growing radical Islamic influence.  Allowing more emigration could help alleviate these pressures in Central Asian countries.  Russia would benefit from migration as a means of compensating for its loss of approximately one million people a year through 2020.  Russia, however, has little experience in integrating migrants from other cultures; Russian nationalism is on the increase as a result of growing ethnic unrest domestically, and our experts believe any efforts to expand immigration policies would be exploited by nationalist politicians.

Ironically, the experts foresaw more unity if economic conditions worsen globally and Eurasia is isolated.  In that case, a stagnant Russia would be looked to by the others to maintain order along the southern rim as some Central Asian countries—Turkmenistan, Tajikistan, and Kyrgyzstan—faced potential collapse.

with the Communist Party retaining control over the central government.

- Younger Chinese leaders who are already exerting influence as mayors and regional officials have been trained in Western-style universities and have a good understanding of international standards of governance.

- Most of the experts at our regional conference, however, believe present and future leaders are agnostic on the issue of democracy and are more interested in developing what they perceive to be the most effective model of governance.

Democratic progress could gain ground in key **Middle Eastern** countries, which thus far have been excluded from the process by repressive regimes. Success in establishing a working democracy in Iraq and Afghanistan—and democratic consolidation in Indonesia—would set an example for other Muslim and Arab states, creating pressures for change.

However, a 2001 Freedom House study showed a dramatic and expanding gap in the levels of freedom and democracy between Islamic countries and the rest of the world. The lack of economic growth in the Middle East outside the energy sector is one of the primary underlying factors for the slow pace. Many regional experts are not hopeful that the generational turnover in several of the regimes will by itself spur democratic reform.

- The extent to which radical Islam grows and how regimes respond to its pressures will also have long-term repercussions for democratization and the growth of civil society institutions, although radicals may use the ballot box to gain power.

- An extended period of high oil prices would allow regimes to put off economic and fiscal reform.

**High-Tech Pressures on Governance.** Today individual PC users have more capacity at their fingertips than NASA had with the computers used in its first moon launches. The trend toward even more capacity, speed, affordability, and mobility will have enormous political implications: myriad individuals and small groups— many of whom had not been previously so empowered—will not only connect with one another but will plan, mobilize, and accomplish tasks with potentially more satisfying and efficient results than their governments can deliver. This almost certainly will affect individuals' relationships with and views of their governments and will put pressure on some governments for more responsiveness.

- China is experiencing among the fastest rates of increase of Internet and mobile phone users in the world, according to the International Telecommunications Union, and is the leading market for broadband communication.

- Reports of growing investment by many Middle Eastern governments in developing high-speed information infrastructures, although they are not yet widely available to the population nor well-connected to the larger world, show obvious potential for the spread of democratic—and undemocratic— ideas.

**Climate Change and Its Implications Through 2020**

Policies regarding climate change are likely to feature significantly in multilateral relations, and the United States, in particular, is likely to face significant bilateral pressure to change its domestic environmental policies and to be a leader in global environmental efforts. There is a strong consensus in the scientific community that the greenhouse effect is real and that average surface temperatures have risen over the last century, but uncertainty exists about causation and possible remedies. Experts in a NIC-sponsored conference judged that concerns about greenhouse gases, of which China and India are large producers, will increase steadily through 2020. There are likely to be numerous weather-related events that, correctly or not, will be linked to global warming. Any of these events could lead to widespread calls for the United States, as the largest producer of greenhouse gases, to take dramatic steps to reduce its consumption of fossil fuels.

Policymakers will face a dilemma: an environmental regime based solely on economic incentives will probably not produce needed technological advances because firms will be hesitant to invest in research when there is great uncertainty about potential profits. On the other hand, a regime based on government regulation will tend to be costly and inflexible. The numerous obstacles to multilateral action include resistance from OPEC countries that depend on fossil fuel revenues, the developing world's view that climate change is a problem created by the industrial world and one they cannot address given their economic constraints, and the need for significant technological innovation to maximize energy efficiency.

Among reasons for optimism, participants noted that the world is ready and eager for US leadership and that new multilateral institutions are not needed to address this issue. Indeed, crafting a policy to limit carbon emissions would be simplified by the fact that three political entities—the United States, the European Union, and China— account for over half of all $CO_2$ emitted into the atmosphere. An agreement that included these three plus the Russian Federation, Japan, and India would cover two-thirds of all carbon emissions.

- Some states will seek to control the Internet and its contents, but they will face increasing challenges as new networks offer multiple means of communicating.

Growing connectivity also will be accompanied by the proliferation of transnational virtual communities of interest, a trend which may complicate the ability of state and global institutions to generate internal consensus and enforce decisions and could even challenge their authority and legitimacy. Groups based on common religious, cultural, ethnic or other affiliations may be torn between their national loyalties and other identities. The potential is considerable for such groups to drive national and even global political decisionmaking on a wide range of issues normally the purview of governments.

The Internet in particular will spur the creation of global movements, which may emerge even more as a robust force in international affairs. For example, technology-enabled diaspora communications in native languages could lead to the preservation of language and culture in the face of widespread emigration and cultural change as well as the generation of political and economic power.

***Populist themes*** are likely to emerge as a potent political and social force, especially as globalization risks aggravating social divisions along economic and ethnic lines. In parts of Latin America particularly, the failure of elites to adapt to the evolving demands of free markets and democracy probably will fuel a revival in populism and drive indigenous movements, which so far have sought change through democratic means, to consider more drastic means for seeking what they consider their "fair share" of political power and wealth.

- However, as with religion, populism will not necessarily be inimical to political development and can serve to broaden participation in the political process. Few experts fear a general backsliding to the rule of military juntas in Latin America.

The Latin American countries that are adapting to challenges most effectively are building sturdier and more capable democratic institutions to implement more inclusive and responsive policies and enhance citizen and investor confidence. A sense of economic progress and hope for its continuance appears essential to the long-term credibility of democratic systems.

Rising nationalism and a trend toward populism also will present a challenge to governments in Asia. Many, such as Laos, Cambodia, and Burma, are unable to deliver on expanding popular demands and risk becoming state failures.

**Latin America in 2020: Will Globalization Cause the Region to Split?**

The experts we consulted in Latin America contended that global changes over the next 15 years could deepen divisions and serve to split Latin America apart in economic, investment, and trade policy terms. As the Southern Cone, particularly Brazil and Chile, reach out to new partners in Asia and Europe, Central America and Mexico, along with Andean countries, could lag behind and remain dependent on the US and Canada as their preferred trade partners and aid providers.

For Latin Americans, government ineffectiveness, in part, prevented many countries from realizing the full measure of economic and social benefits from greater integration into the global economy in the past decade. Instead, the gap between rich and the poor, the represented and the excluded, has grown. Over the next 15 years, the effects of continued economic growth and global integration are likely to be uneven and fragmentary. Indeed, regional experts foresee an increasing risk of the rise of charismatic, self-styled populist leaders, historically common in the region, who would play on popular concerns over inequities between "haves" and "have-nots" in the weakest states in Central America and Andean countries, along with parts of Mexico. In the most profoundly weak of these governments, particularly where the criminalization of the society, and even the state, is most apparent, the leaders could have an autocratic bent and be more stridently anti-American.

The experts made the following observations on regional prospects in other areas:

- **Identity politics.** Increasing portions of the population are identifying themselves as indigenous peoples and will demand not only a voice but, potentially, a new social contract. Many reject globalization as it has played out in the region, viewing it as an homogenizing force that undermines their unique cultures and as a US-imposed, neo-liberal economic model whose inequitably distributed fruits are rooted in the exploitation of labor and the environment.

- **Information technology.** The universalization of the Internet, both as a mass media and means of inter-personal communication, will help educate, connect, mobilize, and empower those traditionally excluded.

- Experts note that a new generation of leaders is emerging in Africa from the private sector; these leaders are much more comfortable with democracy than their predecessors and might provide a strong internal dynamic for democracy in the future.

## Identity Politics

Part of the pressure on governance will come from new forms of identity politics centered on religious convictions and ethnic affiliation. Over the next 15 years, religious identity is likely to become an increasingly important factor in how people define themselves. The trend toward identity politics is linked to increased mobility, growing diversity of hostile groups within states, and the diffusion of modern communications technologies.

- The primacy of ethnic and religious identities will provide followers with a ready-made community that serves as a "social safety net" in times of need—particularly important to migrants. Such communities also provide networks that can lead to job opportunities.

*"Over the next 15 years, religious identity is likely to become an increasingly important factor in how people define themselves."*

While we do not have comprehensive data on the number of people who have joined a religious faith or converted from one faith to another in recent years, trends seem to point toward growing numbers of converts and a *deepening religious commitment* by many religious adherents.

- For example, Christianity, Buddhism, and other religions and practices are spreading in such countries as China as Marxism declines, and the proportion of evangelical converts in traditionally heavily Catholic Latin America is rising.

- By 2020, China and Nigeria will have some of the largest Christian communities in the world, a shift that will reshape the traditionally Western-based Christian institutions, giving them more of an African or Asian or, more broadly, a developing world face.

- Western Europe stands apart from this growing global "religiosity" except for the migrant communities from Africa and the Middle East. Many of the churches' traditional functions—education, social services, etc.—are now performed by the state. A more pervasive, insistent secularism, however, might not foster the cultural acceptance of new Muslim immigrants who view as discriminatory the ban in some West European countries against displays of religious adherence.

Many religious adherents—whether Hindu nationalists, Christian evangelicals in Latin America, Jewish fundamentalists in Israel, or Muslim radicals—are becoming "*activists*." They have a worldview that advocates change of society, a tendency toward making sharp Manichaean distinctions between good and evil, and a religious belief system that connects local conflicts to a larger struggle.

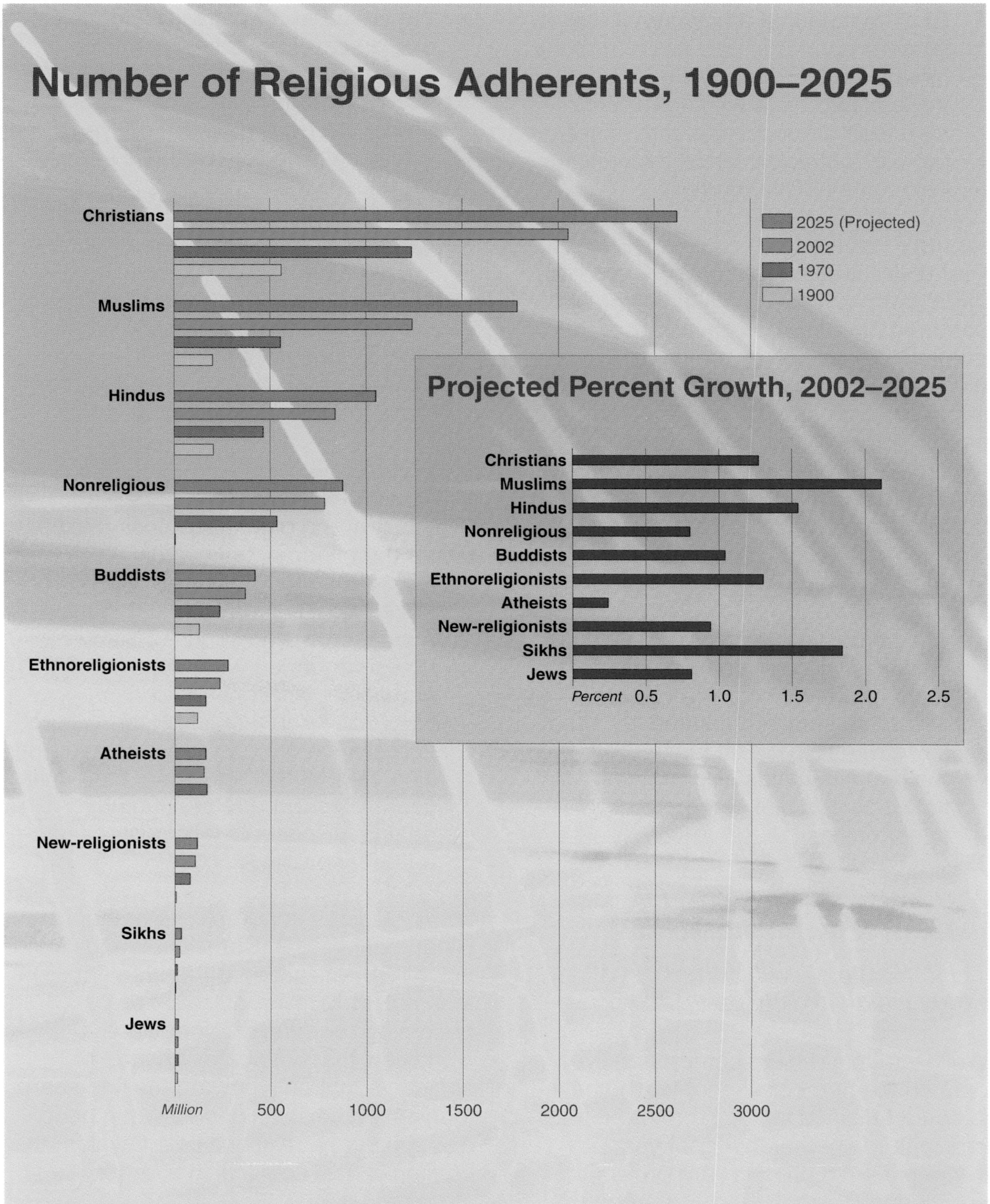

Number of Religious Adherents, 1900–2025
Christians
Muslims
Hindus
Nonreligious
Buddists
Ethnoreligionists
Atheists
New-religionists
Sikhs
Jews
Million
500
1000
1500
2000
2500
3000
2025 (Projected)
2002
1970
1900
Projected Percent Growth, 2002–2025
Christians
Muslims
Hindus
Nonreligious
Buddists
Ethnoreligionists
Atheists
New-religionists
Sikhs
Jews
Percent
0.5
1.0
1.5
2.0
2.5


Source: *Status and Trends in Global Mission as Revealed by the Anual Christian Megacensus, AD 1800–AD 2025*, World Evangelization Research Center.

Such religious-based movements have been common in times of social and political turmoil in the past and have oftentimes been a force for positive change. For example, scholars see the growth of evangelism in Latin America as providing the uprooted, racially disadvantaged and often poorest groups, including women, "with a social network that would otherwise be lacking… providing members with skills they need to survive in a rapidly developing society…(and helping) to promote the development of civil society in the region."[10]

At the same time, the desire by activist groups to change society often leads to more social and political turmoil, some of it violent. In particular, there are likely to be frictions in mixed communities as the activists attempt to gain converts among other religious groups or older established religious institutions. In keeping with the intense religious convictions of many of these movements, activists define their identities in opposition to "outsiders," which can foster strife.

*Radical Islam.* Most of the regions that will experience gains in religious "activists" also have youth bulges, which experts have correlated with high numbers of *radical* adherents, including Muslim extremists.[11]

- Youth bulges are expected to be especially acute in most Middle Eastern and West African countries until at least 2005-2010, and the effects will linger long after.

- In the Middle East, radical Islam's increasing hold reflects the political and economic alienation of many young Muslims from their unresponsive and unrepresentative governments and related failure of many predominantly Muslim states to reap significant economic gains from globalization.

The spread of radical Islam will have a significant global impact leading to 2020, rallying disparate ethnic and national groups and perhaps even creating an authority that transcends national boundaries. Part of the appeal of radical Islam involves its call for a return by Muslims to earlier roots when Islamic civilization was at the forefront of global change. The collective feelings of alienation and estrangement which radical Islam draws upon are unlikely to dissipate until the Muslim world again appears to be more fully integrated into the world economy.

> *"Radical Islam will have a significant global impact… rallying disparate ethnic and national groups and perhaps even creating an authority that trancends national boundaries."*

Radical Islam will continue to appeal to many Muslim migrants who are attracted to the more prosperous West for employment opportunities but do not feel at home in what they perceive as an alien culture.

---

[10] Philip Jenkins, consultations with the National Intelligence Council, August 4, 2004.

[11] We define *Muslim extremists* as a subset of Islamic activists. They are committed to restructuring political society in accordance with their vision of Islamic law and are willing to use violence.

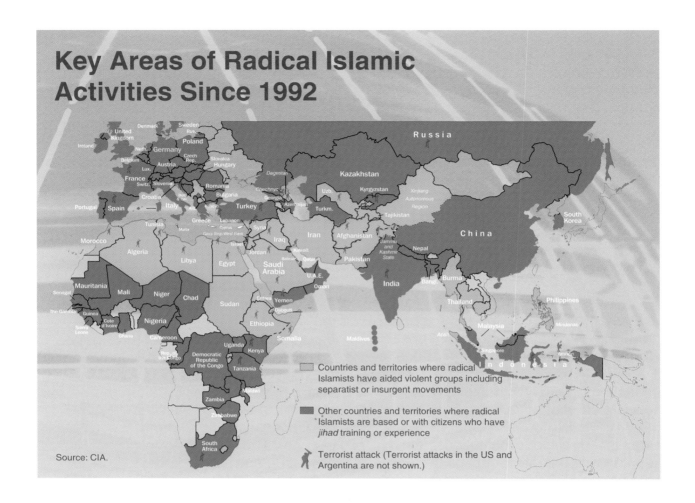

## Key Areas of Radical Islamic Activities Since 1992

Countries and territories where radical Islamists have aided violent groups including separatist or insurgent movements

Other countries and territories where radical Islamists are based or with citizens who have *jihad* training or experience

Terrorist attack (Terrorist attacks in the US and Argentina are not shown.)

Source: CIA.

Studies show that Muslim immigrants are being integrated as West European countries become more inclusive, but many second- and third-generation immigrants are drawn to radical Islam as they encounter obstacles to full integration and barriers to what they consider to be normal religious practices.

Differences over religion and ethnicity also will contribute to future conflict, and, if unchecked, will be a cause of regional strife. Regions where frictions risk developing into wider civil conflict include Southeast Asia, where the historic Christian-Muslim faultlines cut across several countries, including West Africa, The Philippines, and Indonesia.

- Schisms within religions, however historic and longlasting, also could lead to conflict in this era of increased religious identity. A Shia-dominated Iraq is likely to encourage greater activism by Shia minorities in other Middle Eastern nations, such as Saudi Arabia and Pakistan.

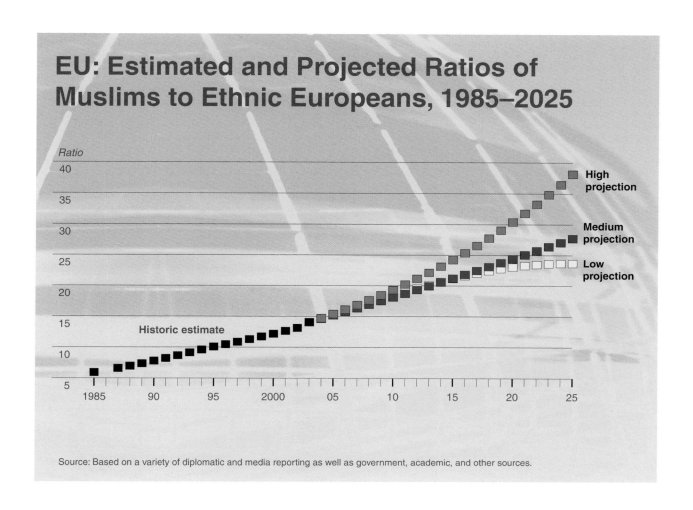

**EU: Estimated and Projected Ratios of Muslims to Ethnic Europeans, 1985–2025**

Ratio

40 — High projection

35

30 — Medium projection

25 — Low projection

20

15 — Historic estimate

10

5

1985  90  95  2000  05  10  15  20  25

Source: Based on a variety of diplomatic and media reporting as well as government, academic, and other sources.

## Fictional Scenario: A New Caliphate

The fictional scenario portrayed below provides an example of how a global movement fueled by radical religious identity could emerge. Under this scenario, a new Caliphate is proclaimed and manages to advance a powerful counter ideology that has widespread appeal. It is depicted in the form of a hypothetical letter from a fictional grandson of Bin Ladin to a family relative in 2020. He recounts the struggles of the Caliph in trying to wrest control from traditional regimes and the conflict and confusion which ensue both within the Muslim world and outside between Muslims and the United States, Europe, Russia and China. While the Caliph's success in mobilizing support varies, places far outside the Muslim core in the Middle East—in Africa and Asia—are convulsed as a result of his appeals. The scenario ends before the Caliph is able to establish both spiritual and temporal authority over a territory— which historically has been the case for previous Caliphates. At the end of the scenario, we identify lessons to be drawn.

# Sa'id Muhammad Bin Ladin

### In the name of God, The Beneficent, The Merciful

June 3, 2020

Grandfather would have been frustrated. The proclamation of the Caliphate has not yet turned out to be our Deliverance.  As you know, dear brother, Grandfather believed in the return to the period of the Rightly Guided Caliphs when the leaders of Islam ruled over an empire as true Defenders of the Faith.  He envisaged the Caliphate holding sway again over the Muslim world, reconquering lost lands in Palestine and Asia, and rooting out the infidel Western influences or "globalization" as the Crusaders so euphemistically call it.  The spiritual and temporal world would once again be in single obedience to the will of Allah, refuting the West's division of Church and State.  In the end, as we've seen, the proclamation has not yet overcome these divisions even if the emergence of the Caliphate has put the fear of Allah in the Crusader powers (and Grandfather would have been pleased at this).  Westernization has certainly lost its luster with many Muslims, and the Caliphate has rent asunder a number of contrived nation-states that were figments of the colonizers' imagination.

When I think back on it, I don't know how we missed seeing the emergence of the young Khalifah—but then we were all surprised, believer and infidel. The young preacher suddenly had a worldwide following. Even before he was proclaimed Khalifah, successor to the Prophet peace be upon Him, he was revered everywhere among the faithful. Maybe it was because he was not al-Qa'ida and had not led a political movement like Grandfather. He appeared all the more spiritual. He wasn't tainted with the unfortunate killings of innocents which, whether Grandfather admitted it or not, deterred some from supporting al-Qa'ida. From Muslim lands halfway around the world in The Philippines, Indonesia, Malaysia, Uzbekistan, Afghanistan and Pakistan, the expressions of allegiance and money came pouring in. Some in the ruling elites also embraced him, hoping to bolster their shaky grip on power. In Europe and America, non-practicing Muslims were awakened to their true identity and faith and, in some cases, left their bewildered Westernized parents to return to their homelands. Even some infidels were impressed with his spirituality; the Pope, for example, tried to initiate an interfaith dialogue with him. And anti-globalizers in the West idolized him. Within a short period, it became clear there was no alternative but to proclaim what so many had yearned for—a new Caliphate.

Oh, what confusion did we sow with the Crusaders. An almost forgotten word reentered the Western lexicon and histories of early Caliphs suddenly rose to be bestsellers on Amazon.com. They had been so smug

*Sa'id Muhammad Bin Ladin*
*In the name of God, The Beneficent, The Merciful*

thinking we had to trudge the same well-worn path behind them toward

secularism if not outright conversion to the so-called Judeo-Christian

value system.  Can you imagine the look on their faces as Muslim

athletes at the Olympics eschewed their national loyalties and instead

proclaimed their allegiance to the Caliphate?  All was being brought

down: the very structures that the West had wrought to imprison us in

their world view—democracy, nation-states, and an international system

run by them—appeared to be in tatters.

Oil was also on their minds and we had them over a barrel like never

before.  As uncertainty raged in the marketplace, rumors circulated about

the US or NATO seizing the oil fields to secure them against a takeover

by the Caliphate.  We heard later that the US could not get its allies to

go along with military intervention and Washington was itself worried

about creating a worldwide Muslim backlash.

At this point, violence with the Shia started—somewhat suspiciously,

increasing the confusion.  The Caliph suspected Iran was stirring up

trouble.  Iran had been upset ever since the Caliphate was proclaimed.

Saudi's Eastern Province with its large Shia population and where the oil

fields are located was particularly vulnerable.  The Gulf state rulers also

played their part.

Sa'id Muhammad Bin Ladin

In the name of God, The Beneficent, The Merciful

And then we suspected Shia-dominated Iraq and behind Iraq the US and the CIA of fomenting the trouble. But if the American infidels were behind it—which I believe was the case—they suffered the consequences. The tenuous peace inside Iraq that America had stitched together so laboriously came undone with the sudden re-igniting of the Sunni insurgency; the insurgents proclaimed themselves the true Caliphate and battled anew both Shia and the American garrisons.

This has been a difficult challenge for the Caliph. The Caliphate represents wholeness and we risk seeing the unraveling of the Caliphate if the age-old Sunni-Shia division cannot be overcome. As a Sunni Arab, I should not show any compassion for a division the partisans of Ali brought upon the Dar al-Islam centuries ago, but, as then, not dealing with the breach has turned out to be a mistake. Some of their new thinkers are joining the likeminded reformers among Sunnis and winning the praise and support of the infidels. The reestablishment of cooperative—even if not warm—relations between the Persians and the Americans would be dangerous for us.

Our people are also too tempted and seduced by Western materialism. Oh the internet—both salvation and a trap set by the devil. It is what is bringing the Caliph so near to total power, spreading his appeal widely.

Sa'id Muhammad Bin Ladin
In the name of God, The Beneficent, The Merciful

But it is also a weapon wielded by our enemies.  More and more of the faithful see how others in the West live and they want those same things, not understanding the vice that accompanies it.

That along with the Sunni-Shia violence flusters the middle class.  They have begun to try to leave our sacred lands.  This has had the ironical result of distressing the Crusader powers in Europe and America.  They do not mind seeing the Caliphate in trouble, but accepting a million or more refugees—most of them accustomed to a high standard of living—has been another thing.  They are on the horns of a dilemma—more Muslims in their midst could stir up resentment; not accepting them could once again underline their hypocrisy.

Confusion reigns elsewhere.  When the Caliphate was proclaimed, the expectation was that regimes would topple immediately.  But that has not happened yet.  Instead there are pockets of followers, undermining many states but not yet succeeding in toppling regimes.  We are very close in Central Asia and in parts of Pakistan and Afghanistan where a civil war rages.  Russia is bogged down in fighting insurgencies and propping up Central Asian autocracies.  In a paradoxical twist the United States is Russia's ally this time:  instead of supplying money and arms for the *mujahidin* to fight the Soviets, they are helping the Russians to

Sa'id Muhammad Bin Ladin
In the name of God, The Beneficent, The Merciful

fight the *mujahidin*. In some countries two sets of authorities rule, neither of which has full control. Pashtunistan is declared but has not yet fully established its power. All of this has begun to deter the Crusaders from exploiting our resources. No pipelines are now being built and some were put out of business. Rising Asia has been hurt because construction of new pipelines has come to a halt. China eschewed any help from the US but has become increasingly concerned about Muslim bourgeois "irredentism" in its midst.

Southeast Asia and parts of East and West Africa are going almost the same way. The proclamation of the Caliphate emboldened the insurgents but has so far had its most powerful effect in driving a wedge into states and creating pockets where the Caliphate and Sharia rule. The Asians are difficult, though. They think they have discovered an Asian way.

The struggle continues in Palestine too, despite the creation of a Palestinian state. The Zionists still occupy Muslim lands and share control of Jerusalem. The Europeans thought they could dodge a clash of civilizations, but they now see that growing numbers of Muslims in their midst are turning to the Caliph. The US also has infidel followers of the Caliph, including a senator's daughter. The US is caught in the middle, trying to buy off those supporting the Caliphate in order to

Sa'id Muhammad Bin Ladin
*In the name of God, The Beneficent, The Merciful*

pacify the Iraqi Sunnis, but not wanting to alienate Israel. Europe is piling the pressure on the Zionists for the first time, talking of sanctions against Israel.

So much confusion and turmoil, but this is great for us. At first with the new Caliphate our numbers dwindled and the core al-Qa'ida went out of business, but then began so many new struggles. We can fight to regain Muslim lands, even if the Caliph has turned its back on some struggles and is in danger of becoming soft. We cut deals with the local warlords, exploiting their hospitality or paying tribute and are largely left free to operate as we see fit. I'm hopeful...

## "Lessons Learned"

A Caliphate would not have to be entirely successful for it to present a serious challenge to the international order. This scenario underlines the saliency of the cross-cultural ideological debate that would intensify with growing religious identities.

- The IT revolution is likely to amplify the clash between Western and Muslim worlds.

- The appeal of a Caliphate among Muslims would vary from region to region, which argues for Western countries adopting a differentiated approach to counter it. Muslims in regions benefiting from globalization, such as parts of Asia and Europe, may be torn between the idea of a spiritual Caliphate and the material advantages of a globalized world.

- The proclamation of a Caliphate would not lessen the likelihood of terrorism and, in fomenting more conflict, could fuel a new generation of terrorists intent on attacking those opposed to the Caliphate, whether inside or outside the Muslim world.

# Pervasive Insecurity

We foresee a more pervasive sense of insecurity, which may be as much based on psychological perceptions as physical threats, by 2020. The psychological aspects, which we have addressed earlier in this paper, include concerns over job security as well as fears revolving around migration among both host populations and migrants.

Terrorism and internal conflicts could interrupt the process of globalization by significantly increasing the security costs associated with international commerce, encouraging restrictive border control policies, and adversely affecting trade patterns and financial markets. Although far less likely than internal conflicts, conflict among great powers would create risks to world security. The potential for the proliferation of weapons of mass destruction (WMD) will add to the pervasive sense of insecurity.

**Transmuting International Terrorism**
The key factors that spawned international terrorism show no signs of abating over the next 15 years. Experts assess that the majority of international terrorist groups will continue to identify with radical Islam. The revival of Muslim identity will create a framework for the spread of radical Islamic ideology both inside and outside the Middle East, including Western Europe, Southeast Asia and Central Asia.

- This revival has been accompanied by a deepening solidarity among Muslims caught up in national or regional separatist struggles, such as Palestine, Chechnya, Iraq, Kashmir, Mindanao, or southern Thailand and has emerged in response to government repression, corruption, and ineffectiveness.

- A radical takeover in a Muslim country in the Middle East could spur the spread of terrorism in the region and give confidence to others that a new Caliphate is not just a dream.

- Informal networks of charitable foundations, *madrasas, hawalas,*[12] and other mechanisms will continue to proliferate and be exploited by radical elements.

- Alienation among unemployed youths will swell the ranks of those vulnerable to terrorist recruitment.

*"Our greatest concern is that [terrorist groups] might acquire biological agents, or less likely, a nuclear device, either of which could cause mass casualties."*

There are indications that the Islamic radicals' professed desire to create a transnational insurgency, that is, a drive by Muslim extremists to overthrow a number of allegedly apostate secular

---

[12] *Hawalas* constitute an informal banking system.

governments with predominately Muslim subjects, will have an appeal to many Muslims.

- Anti-globalization and opposition to US policies could cement a greater body of terrorist sympathizers, financiers, and collaborators.

*"…We expect that by 2020 al-Qa'ida will have been superceded by similarly inspired but more diffuse Islamic extremist groups."*

**A Dispersed Set of Actors.** Pressure from the global counterterrorism effort, together with the impact of advances in information technology, will cause the terrorist threat to become increasingly decentralized, evolving into an eclectic array of groups, cells, and individuals. While taking advantage of sanctuaries around the world to train, terrorists will not need a stationary headquarters to plan and carry out operations. Training materials, targeting guidance, weapons know-how, and fund-raising will increasingly become virtual (i.e., online).

The core al-Qa'ida membership probably will continue to dwindle, but other groups inspired by al-Qa'ida, regionally based groups, and individuals labeled simply as jihadists—united by a common hatred of moderate regimes and the West—are likely to conduct terrorist attacks. The al-Qa'ida membership that was distinguished by having trained in Afghanistan will gradually dissipate, to be replaced in part by the dispersion of the experienced survivors of the conflict in Iraq. We expect that by 2020 al-Qa'ida will have been superceded by similarly

inspired but more diffuse Islamic extremist groups, all of which will oppose the spread of many aspects of globalization into traditional Islamic societies.

- Iraq and other possible conflicts in the future could provide recruitment, training grounds, technical skills and language proficiency for a new class of terrorists who are "professionalized" and for whom political violence becomes an end in itself.

- Foreign jihadists—individuals ready to fight anywhere they believe Muslim lands are under attack by what they see as "infidel invaders"—enjoy a growing sense of support from Muslims who are not necessarily supporters of terrorism.

Even if the number of extremists dwindles, however, the terrorist threat is likely to remain. Through the Internet and other wireless communications technologies, individuals with ill intent will be able to rally adherents quickly on a broader, even global, scale and do so obscurely. The rapid dispersion of bio- and other lethal forms of technology increases the potential for an individual not affiliated with any terrorist group to be able to inflict widespread loss of life.

**Weapons, Tactics, and Targets.** In the past, terrorist organizations relied on state sponsors for training, weapons, logistical support, travel documents, and money in support of their operations. In a globalized world, groups such as Hizballah are increasingly self-sufficient in meeting these needs and may act in a state-like manner to preserve "plausible deniability" by supplying other groups, working through third parties to meet their objectives, and even engaging governments diplomatically.

Most terrorist attacks will continue to employ primarily conventional weapons, incorporating new twists to keep counterterrorist planners off balance. Terrorists probably will be most original not in the technologies or weapons they employ but rather in their operational concepts—i.e., the scope, design, or support arrangements for attacks.

- One such concept that is likely to continue is a large number of simultaneous attacks, possibly in widely separated locations.

While vehicle-borne improvised explosive devices will remain popular as asymmetric weapons, terrorists are likely to move up the technology ladder to employ advanced explosives and unmanned aerial vehicles.

*"Terrorist use of biological agents is therefore likely, and the range of options will grow."*

The religious zeal of extremist Muslim terrorists increases their desire to perpetrate attacks resulting in high casualties. Historically, religiously inspired terrorism has been most destructive because such groups are bound by few constraints.

The most worrisome trend has been an intensified search by some terrorist groups to obtain weapons of mass destruction. Our greatest concern is that these groups might acquire biological agents or less likely, a nuclear device, either of which could cause mass casualties.

- Bioterrorism appears particularly suited to the smaller, better-informed groups. Indeed, the bioterrorist's laboratory could well be the size of a household kitchen, and the weapon built there could be smaller than a toaster. Terrorist use of biological agents is therefore likely, and the range of options will grow. Because the recognition of anthrax, smallpox or other diseases is typically delayed, under a "nightmare scenario" an attack could be well under way before authorities would be cognizant of it.

- The use of radiological dispersal devices can be effective in creating panic because of the public's misconception of the capacity of such attacks to kill large numbers of people.

With advances in the design of simplified nuclear weapons, terrorists will continue to seek to acquire fissile material in order to construct a nuclear weapon. Concurrently, they can be expected to continue attempting to purchase or steal a weapon, particularly in Russia or Pakistan. Given the possibility that terrorists could acquire nuclear weapons, the use of such weapons by extremists before 2020 cannot be ruled out.

We expect that terrorists also will try to acquire and develop the capabilities to conduct cyber attacks to cause physical damage to computer systems and to disrupt critical information networks.

The United States and its interests abroad will remain prime terrorist targets, but more terrorist attacks might

**Organized Crime**

Changing geostrategic patterns will shape global organized criminal activity over the next 15 years. Organized crime is likely to thrive in resource-rich states undergoing significant political and economic transformation, such as India, China, Russia, Nigeria, and Brazil as well as Cuba, if it sees the end of its one-party system. Some of the former states of the Soviet Union and the Warsaw Pact also will remain vulnerable to high levels of organized crime.

- States that transition to one-party systems—such as any new Islamic-run state—will be vulnerable to corruption and attendant organized crime, particularly if their ideology calls for substantial government involvement in the economy.

- Changing patterns of migration may introduce some types of organized crime into countries that have not previously experienced it. Ethnic-based organized crime groups typically prey on members of their own diasporas and use them to gain footholds in new regions.

Some organized crime syndicates will form loose alliances with one another. They will attempt to corrupt leaders of unstable, economically fragile, or failing states, insinuate themselves into troubled banks and businesses, exploit information technologies, and cooperate with insurgent movements to control substantial geographic areas.

Organized crime groups usually do not want to see governments toppled but thrive in countries where governments are weak, vulnerable to corruption, and unable or unwilling to consistently enforce the rule of law.

- Criminal syndicates, particularly drug trafficking syndicates, may take virtual control of regions within failing states to which the central government cannot extend its writ.

If governments in countries with WMD capabilities lose control of their inventories, the risk of organized crime trafficking in nuclear, biological, or chemical weapons will increase between now and 2020.

We expect that the relationship between terrorists and organized criminals will remain primarily a matter of business, i.e., that terrorists will turn to criminals who can provide forged documents, smuggled weapons, or clandestine travel assistance when the terrorists cannot procure these goods and services on their own. Organized criminal groups, however, are unlikely to form long-term strategic alliances with terrorists. Organized crime is motivated by the desire to make money and tends to regard any activity beyond that required to effect profit as bad for business. For their part, terrorist leaders are concerned that ties to non-ideological partners will increase the chance of successful police penetration or that profits will seduce the faithful.

encounter the most severe and most frequent outbreaks of violence. For the most part, those states most susceptible to violence are in a great arc of instability from Sub-Saharan Africa, through North Africa, into the Middle East, the Balkans, the Caucasus and South and Central Asia and through parts of Southeast Asia. Countries in these regions are generally those "behind" the globalization curve.

- The number of internal conflicts is down significantly since the late 1980s and early 1990s, when the breakup of the Soviet Union and Communist regimes in Central Europe allowed suppressed ethnic and nationalist strife to flare. Although a leveling off point has been reached, the continued prevalence of troubled and institutionally weak states creates conditions for such conflicts to occur in the future.

*"Lagging economies, ethnic affiliations, intense religious convictions, and youth bulges will align to create a 'perfect storm' [for] internal conflict."*

be aimed at Middle Eastern regimes and at Western Europe.

**Intensifying Internal Conflicts**
Lagging economies, ethnic affiliations, intense religious convictions, and youth bulges will align to create a "perfect storm," creating conditions likely to spawn internal conflict. The governing capacity of states, however, will determine whether and to what extent conflicts actually occur. Those states unable both to satisfy the expectations of their peoples and to resolve or quell conflicting demands among them are likely to

Internal conflicts are often particularly vicious, long-lasting, and difficult to terminate. Many of these conflicts generate internal displacements and external refugee flows, destabilizing neighboring countries.

- Sub-Saharan Africa will continue to be particularly at risk for major new or worsening humanitarian emergencies stemming from conflict. Genocidal conflicts aimed at annihilating all or part of a racial, religious, or ethnic group, and conflicts caused by other crimes against humanity—such as

97

forced, large-scale expulsions of populations—are particularly likely to generate migration and massive, intractable humanitarian needs.

*"Africa in 2020 ... will increasingly resemble a patchwork quilt with significant differences in economic and political performance."*

Some internal conflicts, particularly those that involve ethnic groups straddling national boundaries, risk escalating into regional conflicts. At their most extreme, internal conflicts can produce a failing or failed state, with expanses of territory and populations devoid of effective governmental control. In such instances, those territories can become sanctuaries for transnational terrorists (like al-Qa'ida in Afghanistan) or for criminals and drug cartels (such as in Colombia).

**Rising Powers: Tinder for Conflict?**
The likelihood of great power conflict escalating into total war in the next 15 years is lower than at any time in the past century, unlike during previous centuries when local conflicts sparked world wars. The rigidities of alliance systems before World War I and during the interwar period, as well as the two-bloc standoff during the Cold War, virtually assured that small conflicts would be quickly generalized. Now, however, even if conflict would break out over Taiwan or between India and Pakistan, outside powers as well as the primary actors would want to limit its extent. Additionally, the growing dependence on global financial and trade networks increasingly will act as a deterrent to conflict among the great powers—the US, Europe, China, India, Japan and Russia.

This does not eliminate the possibility of great power conflict, however. The absence of effective conflict resolution mechanisms in some regions, the rise of nationalism in some states, and the raw emotions on both sides of key issues increase the chances for miscalculation.

- Although a military confrontation between China and Taiwan would derail Beijing's efforts to gain acceptance as a regional and global power, we cannot discount such a possibility. Events such as Taiwan's proclamation of independence could lead Beijing to take steps it otherwise might want to avoid, just as China's military buildup enabling it to bring overwhelming force against Taiwan increases the risk of military conflict.

- India and Pakistan appear to understand the likely prices to be paid by triggering a conflict. But nationalistic feelings run high and are not likely to abate. Under plausible scenarios Pakistan might use nuclear weapons to counter success by the larger Indian conventional forces, particularly given Pakistan's lack of strategic depth.

*"Advances in modern weaponry—longer ranges, precision delivery, and more destructive conventional munitions—create circumstances encouraging the preemptive use of military force."*

Should conflict occur that involved one or more of the great powers, the consequences would be significant. Advances in modern weaponry—longer

## How Can Sub-Saharan Africa Move Forward?

Most of the regional experts we consulted believe the most likely scenario for Africa in 2020 is that it will increasingly resemble a patchwork quilt with significant differences in economic and political performance.

Africa's capacity to benefit from the positive elements of globalization will depend on the extent to which individual countries can bring an end to conflict, improve governance, rein in corruption, and establish the rule of law. If progress is achieved in these areas, an expansion of foreign investment, which currently is mostly confined to the oil sector, is possible. Our regional experts felt that if African leaders used such investment to help their economies grow—opening avenues to wealth other than through the power of the state—they might be able to mitigate the myriad other problems facing their countries, with the prospect of prosperity decreasing the level of conflict.

Expanded development of existing or new sources of wealth will remain key. Although mineral and natural resources are not evenly distributed among its countries, Sub-Saharan Africa is well endowed with them and has the potential not only to be self-sufficient in food, but to become a major exporter of agricultural, animal, and fish products. The lowering or elimination of tariff barriers and agricultural subsidies in the European Union and the United States, combined with the demand for raw materials from the burgeoning Chinese and Indian economies, could provide major stimulus to African economies and overcome decades of depressed commodity prices.

African experts have agreed that economic reform and good governance are essential for high economic growth and also have concluded that African countries must take the initiative in negotiating new aid and trade relationships that heretofore were essentially dictated by the international financial institutions and the developed world. The New Partnership for Africa's Development (NEPAD), with its peer review mechanism, provides one mechanism for bringing about this economic transformation, if its members individually and collectively honor their commitments.

Over the next 15 years, democratic reform will remain slow and imperfect in many countries due to a host of social and economic problems, but it is highly unlikely that democracy will be challenged as the norm in Africa. African leaders face alliances of international and domestic nongovernmental organizations that sometimes want to supplant certain state services, criminal networks that operate freely across borders, and Islamic groups bent on establishing safehavens. Some states may fail but in others the overall quality of democracy probably will increase. An emerging generation of leaders includes many from the private sector, who are more comfortable with democracy than their predecessors and who could provide a strong political dynamic for democracy in the future.

Leadership will remain the ultimate wild card, which, even in the least promising circumstances, could make a huge, positive difference. Although countries with poor leadership will find it harder not to fail, those with good leadership that promotes order, institutions, and conflict resolution will at least have a chance of progressing.

ranges, precision delivery, and more destructive conventional munitions—create circumstances encouraging the preemptive use of military force. The increased range of new missile and aircraft delivery systems provides sanctuary to their possessors.

Until strategic defenses become as strong as strategic offenses, there will be great premiums associated with the ability to expand conflicts geographically in order to deny an attacker sanctuary. Moreover, a number of recent high-technology conflicts have demonstrated that the outcomes of early battles of major conflicts most often determine the success of entire campaigns. Under these circumstances, military experts believe preemption is likely to appear necessary for strategic success.

### The WMD Factor
***Nuclear Weapons.*** Over the next 15 years, a number of countries will continue to pursue their nuclear, chemical, and biological weapons programs and in some cases will enhance their capabilities. Current nuclear weapons states will continue to improve the survivability of their deterrent forces and almost certainly will improve the reliability, accuracy, and lethality of their delivery systems as well as develop capabilities to penetrate missile defenses. The open demonstration of nuclear capabilities by any state would further *discredit the current nonproliferation regime*, cause a possible shift in the balance of power, and increase the risk of conflicts escalating into nuclear ones.

- Countries without nuclear weapons, especially in the Middle East and Northeast Asia, may decide to seek them as it becomes clear that their

neighbors and regional rivals already are doing so.

- The assistance of proliferators, including former private entrepreneurs such as the A.Q. Khan network, will reduce the time required for additional countries to develop nuclear weapons.

*"Countries without nuclear weapons … may decide to seek them as it becomes clear that their neighbors and regional rivals are already doing so."*

***Chemical and Biological Weapons.*** Developments in CW and BW agents and the proliferation of related expertise will pose a substantial threat, particularly from terrorists, as we have noted.

- Given the goal of some terrorist groups to use weapons that can be employed surreptitiously and generate dramatic impact, we expect to see terrorist use of some readily available biological and chemical weapons.

Countries will continue to integrate both CW and BW production capabilities into apparently legitimate commercial infrastructures, further concealing them from scrutiny, and BW/CW programs will be less reliant on foreign suppliers.

- Major advances in the biological sciences and information technology probably will accelerate the pace of BW agent development, increasing the potential for agents that are more difficult to detect or to defend against. Through 2020 some countries will continue to try to develop chemical agents designed to circumvent the

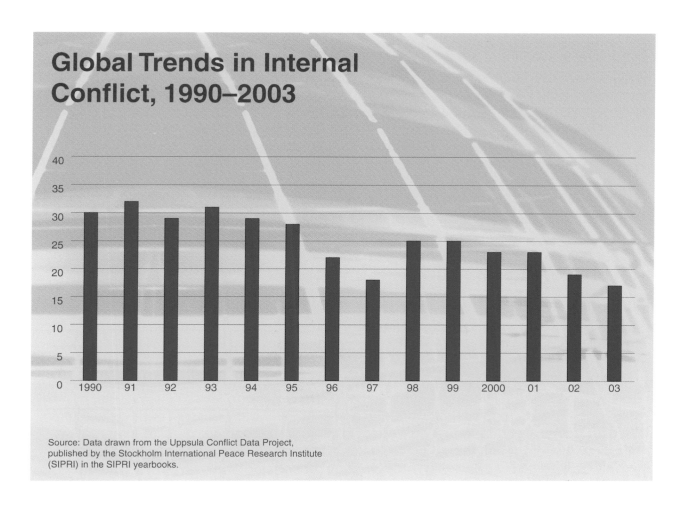

## Global Trends in Internal Conflict, 1990–2003

Source: Data drawn from the Uppsula Conflict Data Project, published by the Stockholm International Peace Research Institute (SIPRI) in the SIPRI yearbooks.

Chemical Weapons Convention verification regime.

*"Developments in CW and BW agents and the proliferation of related expertise will pose a substantial threat, particularly from terrorists..."*

**Delivery Systems.** Security will remain at risk from increasingly advanced and lethal ballistic and cruise missiles and unmanned aerial vehicles (UAVs). States almost certainly will continue to increase the range, reliability, and accuracy of the missile systems in their inventories. By 2020 several countries of concern probably will have acquired Land-Attack Cruise Missiles (LACMs) capable of threatening the US Homeland if brought closer to US shores. Both North Korea and Iran probably will have an ICBM capability well before 2020 and will be working on improvements to enhance such capabilities, although new regimes in either country could rethink these objectives. Several other countries are likely to develop space launch vehicles (SLVs) by 2020 to put domestic satellites in orbit and to enhance national prestige. An SLV is a key stepping-stone toward an ICBM: it could be used as a booster in an ICBM development.

**International Institutions in Crisis**

Increased pressures on international institutions will incapacitate many, unless and until they can be radically adapted to accommodate new actors and new priorities. Regionally based institutions will be particularly challenged to meet the complex transnational threats posed by economic upheavals, terrorism, organized crime, and WMD proliferation. Such post-World War II creations as the United Nations and international financial institutions risk sliding into obsolescence unless they take into consideration the growing power of the rising powers.

- Both supporters and opponents of multilateralism agree that Rwanda, Bosnia, and Somalia demonstrated the ineffectiveness, lack of preparation, and weaknesses of global and regional institutions to deal with what are likely to be the more common types of conflict in the future.

The problem of state failure—which is a source or incubator for a number of transnational threats—argues for better coordination between institutions, including the international financial ones and regional security bodies.

Building a global consensus on how and when to intervene is likely to be the biggest hurdle to greater effectiveness but essential in many experts' eyes if multilateral institutions are to live up to their potential and promise. Many states, especially the emerging powers, continue to worry about setting precedents for outside intervention that can be used against them. Nevertheless, most problems, such as failing states, can only be effectively dealt with through early recognition and preventive measures.

Other issues that are likely to emerge on the international agenda will add to the pressures on the collective international order as well as on individual countries. These "new" issues could become the staples of international diplomacy much as human rights did in the 1970s and 1980s. Ethical issues linked to biotechnological discoveries such as cloning, GMOs, and access to biomedicines could become the source of hot debates among countries and regions. As technology increases the capabilities of states to track terrorists, concerns about privacy and extraterritoriality may increasingly surface among publics worldwide. Similarly, debates over environmental issues connected with tempering climate change risk scrambling the international order, pitting the US against its traditional European allies, as well as developed countries against the developing world, unless more global cooperation is achieved. Rising powers may see in the ethical and environmental debates an attempt by the rich countries to slow down their progress by imposing "Western" standards or values. Institutional reform might increasingly surface as an issue. Many in the developing world believe power in international bodies is too much a snapshot of the post-World War II world rather than the current one.

## The Rules of War: Entering "No Man's Land"

With most armed conflict taking *unconventional* or *irregular* forms—such as humanitarian interventions and operations designed to root out terrorist home bases—rather than conventional state-to-state warfare, the principles covering resort to, and use of, military force will increasingly be called into question. Both the international law enshrining territorial sovereignty and the Geneva Conventions governing the conduct of war were developed before transnational security threats like those of the twenty-first century were envisioned.

In the late 1990s, the outcry over former Serbian President Milosevic's treatment of Kosovars spurred greater acceptance of the principle of international humanitarian interventions, providing support to those in the "just war" tradition who have argued since the founding of the UN and before that the international community has a "duty to intervene" in order to prevent human rights atrocities. This principle, however, continues to be vigorously contested by countries worried about harm to the principle of national sovereignty.

The legal status and rights of prisoners taken during military operations and suspected of involvement in terrorism will be a subject of controversy—as with many captured during Operation ENDURING FREEDOM in Afghanistan. A debate over the degree to which religious leaders and others who are perceived as abetting or inciting violence should be considered international terrorists is also likely to come to the fore.

The Iraq war has raised questions about what kind of status, if any, to accord to the increasing number of contractors used by the US military to provide security, man POW detention centers, and interrogate POWs or detainees.

Protection for nongovernmental organizations (NGOs) in conflict situations is another issue that has become more complicated as some charitable work—such as Wahabi missionaries funding terrorist causes—has received criticism and enforcement action at the same time that Western and other NGOs have become "soft targets" in conflict situations.

The role of the United States in trying to set norms is itself an issue and probably will complicate efforts by the global community to come to an agreement on a new set of rules. Containing and limiting the scale and savagery of conflicts will be aggravated by the absence of clear rules.

> "Such post-World War II creations as the UN and international financial institutions risk sliding into obsolescence unless they take into consideration the growing power of the developing world."

## Post-Combat Environments Pose the Biggest Challenge

For the United States particularly, if the past decades are any guide waging and winning a conventional war is unlikely to be much of a challenge over the next 15 years in light of our overarching capabilities to conduct such a war. However, the international community's efforts to prevent outbreaks and ensure that conflicts are not a prelude to new ones could remain elusive.

- Nation-building is at best an imperfect concept, but more so with the growing importance of cultural, ethnic, and religious identities.

- Africa's effort to build a regional peacekeeping force shows some promise, but Sub-Saharan Africa will struggle with attracting sufficient resources and political will.

- The enormous costs in resources and time for meaningful nation-building or post-conflict/failed state stability operations are likely to be a serious constraint on such coalition or international commitments.

## Fictional Scenario: Cycle of Fear

*This scenario explores what might happen if proliferation concerns increased to the point that large-scale intrusive security measures were taken. In such a world, proliferators—such as illegal arms merchants—might find it increasingly hard to operate, but at the same time, with the spread of WMD, more countries might want to arm themselves for their own protection. This scenario is depicted in a series of text-message exchanges between two arms dealers. One is ideologically committed to leveling the playing field and ensuring the Muslim world has its share of WMD, while the other is strictly for hire. Neither knows for sure who is at the end of his chain—a government client or terrorist front. As the scenario progresses, the cycle of fear originating with WMD-laden terrorist attacks has gotten out of hand—to the benefit of the arms dealers, who appear to be engaged in lucrative deals. However, fear begets fear. The draconian measures increasingly implemented by governments to stem proliferation and guard against terrorism also have the arms dealers beginning to run scared. In all of this, globalization may be the real victim.*

# Policy Implications

The international order will be in greater flux in the period out to 2020 than at any point since the end of the Second World War. As we map the future, the prospects for global prosperity and the limited likelihood of great power conflict provide an overall favorable environment for coping with the challenges ahead. Despite daunting challenges, the United States, in particular, will be better positioned than most countries to adapt to the changing global environment.

As our scenarios illustrate, we see several ways in which major global changes could begin to take shape and be buffeted or bolstered by the forces of change over the next 15 years. In a sense, the scenarios provide us with four different *lenses* on future developments, underlining the wide range of factors, discontinuities, and uncertainties shaping a new global order. One lens is the globalized economy, another is the security role played by the US, a third is the role of social and religious identity, and a fourth is the breakdown of the international order because of growing insecurity. They highlight various "switching points" that could shift developments onto one path or the other. The most important tipping points include the impact of robust economic growth and the spread of technology; the nature and extent of terrorism; the resiliency or weakness of states, particularly in the Middle East, Central Asia, and Africa; and the potential spread of conflict, including between states.

- On balance, for example, as the hypothetical **Davos World** scenario shows, robust economic growth probably will help to overcome divisions and pull more regions and countries into a new global order. However, the rapid changes might also produce disorder at times; one of the lessons of that and the other scenarios is the need for management to ensure globalization does not go off the rails.

The evolving framework of international politics in all the scenarios suggests that **nonstate actors** will continue to assume a more prominent role even though they will not displace the nation-state. Such actors range from terrorists, who will remain a threat to global security, to NGOs and global firms, which exemplify largely positive forces by spreading technology, promoting social and economic progress, and providing humanitarian assistance.

The United States and other countries throughout the world will continue to be vulnerable to **international terrorism.** As we have noted in the **Cycle of Fear** scenario, terrorist campaigns that escalate to unprecedented heights, particularly if they involve WMD, are one of the few developments that could threaten globalization.

Counterterrorism efforts in the years ahead—against a more diverse set of terrorists who are connected more by ideology and technology than by geography—will be a more elusive challenge than focusing on a relatively centralized organization such as al-Qa'ida. The looser the connections

**Is the United States' Technological Prowess at Risk?**

US investment in basic research and the innovative application of technology has directly contributed to US leadership in economic and military power during the post-World War II era. Americans, for example, invented and commercialized the semiconductor, the personal computer, and the Internet with other countries following the US lead.[a] While the United States is still the present leader, there are signs this leadership is at risk.

The number of US engineering graduates peaked in 1985 and is presently down 20 percent from that level; the percentage of US undergraduates taking engineering is the second lowest of all developed countries. China graduates approximately three times as many engineering students as the United States. However, post-9/11 security concerns have made it harder to attract incoming foreign students and, in some cases, foreign nationals available to work for US firms.[b] Non-US universities—for which a US visa is not required—are attempting to exploit the situation and bolster their strength.

Privately funded industrial research and development—which accounts for 60 percent of the US total—while up this year, suffered three previous years of decline.[c] Further, major multinational corporations are establishing corporate "research centers" outside of the United States.

While these signs are ominous, the integrating character of globalization and the inherent strengths of the US economic system preclude a quick judgment of an impending US technological demise. By recent assessments, the United States is still the most competitive society in the world among major economies.[d] In a globalized world where information is rapidly shared—including cross-border sharing done internally by multinational corporations—the creator of new science or technology may not necessarily be the beneficiary in the marketplace.

---

[a] "Is America Losing Its Edge? Innovation in a Globalized World." Adam Segal, *Foreign Affairs*, November December 2004; New York, NY p.2.

[b] "Observations on S&T Trends and Their Potential Impact on Our Future." William Wulf (President, National Academy of Engineering). Paper submitted to the Center for Strategic and International Studies (CSIS) in support of the National Intelligence Council 2020 Study, Summer 2004.

[c] "Is America Losing Its Edge?," p.3.

[d] *Global Competitiveness Report 2004-2005*, World Economic Forum, http://www.weforum.org. October 2004.

among individual terrorists and various cells, the more difficult it will be to uncover and disrupt terrorist plotting.

- One of our scenarios—*Pax Americana*—envisages a case in which US and European consensus on fighting terrorism would grow much stronger but, under other scenarios, including the hypothetical *New Caliphate*, US, Russian, Chinese and European interests diverge, possibly limiting cooperation on counterterrorism.

*"The US will have to battle world public opinion, which has dramatically shifted since the end of the Cold War."*

The success of the US-led global counterterrorism campaign will hinge on the capabilities and resolve of individual countries to fight terrorism on their own soil. Efforts by Washington to bolster the capabilities of local security forces in other countries and to work with them on their priority issues (such as soaring crime) would be likely to increase cooperation.

- Defense of the US Homeland will begin overseas. As it becomes more difficult for terrorists to enter the United States, they are likely to try to attack the Homeland from neighboring countries.

A counterterrorism strategy that approaches the problem on multiple fronts offers the greatest chance of containing—and ultimately reducing—the terrorist threat. The development of more open political systems, broader economic opportunities, and empowerment of Muslim reformers would be viewed positively by the broad Muslim communities who do not support the radical agenda of Islamic extremists. *A New Caliphate* scenario dramatizes the challenge of addressing the underlying causes of extremist violence, not just its manifest actions.

- The Middle East is unlikely to be the only battleground in which this struggle between extremists and reformers occurs. European and other Muslims outside the Middle East have played an important role in the internal ideological conflicts, and the degree to which Muslim minorities feel integrated in European societies is likely to have a bearing on whether they see a clash of civilizations as inevitable or not. Southeast Asia also has been increasingly a theater for terrorism.

Related to the terrorist threat is the problem of the *proliferation of WMD* and the potential for countries to have increased motivation to acquire nuclear weapons if their neighbors and regional rivals are doing so. As illustrated in the *Cycle of Fear* scenario, global efforts to erect greater barriers to the spread of WMD and to dissuade any other countries from seeking nuclear arms or other WMD as protection will continue to be a challenge. As various of our scenarios underline, the communications revolution gives proliferators a certain advantage in striking deals with each other and eluding the authorities, and the "assistance" they provide can cut years off the time it would take for countries to develop nuclear weapons.

## How the World Sees the United States

In the six regional conferences that we hosted we asked participants about their views of the role of the United States as a driver in shaping developments in their regions and globally.

### Asia
Participants felt that US preoccupation with the war on terrorism is largely irrelevant to the security concerns of most Asians. The key question that the United States needs to ask itself is whether it can offer Asian states an appealing vision of regional security and order that will rival and perhaps exceed that offered by China.

US disengagement from what matters to US Asian allies would increase the likelihood that they would climb on Beijing's bandwagon and allow China to create its own regional security order that excludes the United States.

Participants felt that the rise of China need not be incompatible with a US-led international order. The critical question is whether or not the order is flexible enough to adjust to a changing distribution of power on a global level. An inflexible order would increase the likelihood of political conflict between emerging powers and the United States. If the order is flexible, it may be possible to forge an accommodation with rising powers and strengthen the order in the process.

### Sub-Saharan Africa
Sub-Saharan African leaders worry that the United States and other advantaged countries will "pull up the drawbridge" and abandon the region.

Participants opined that the United States and other Western countries may not continue to accept Africa's most successful "export," its people. The new African diaspora is composed overwhelmingly of economic migrants rather than political migrants as in previous eras.

Some participants felt that Africans worry that Western countries will see some African countries as "hopeless" over the next 15 years because of prevailing economic conditions, ecological problems, and political circumstances.

Participants feared that the United States will focus only on those African countries that are successful.

### Latin America
Conference participants acknowledged that the United States is the key economic, political, and military player in the hemisphere. At the same time, Washington was viewed as traditionally not paying sustained attention to the region and, instead of responding to systemic problems, as reacting primarily to crises. Participants saw a fundamentalist trend in Washington that would lead to isolation and unilateralism and undercut cooperation. Most shared the view that the US "war on terrorism" had little to do with Latin America's security concerns.

Latin American migrants are a stabilizing force in relations with the United States. An important part of the US labor pool, migrants also remit home needed dollars along with new views on democratic governance and individual initiative that will have a positive impact on the region.

*(Continued on next page...)*

US policies also can have a positive impact. Some participants said the region would benefit from US application of regional mechanisms to resolve problems rather than punitive measures against regimes not to its liking, such as that of Fidel Castro.

## Middle East

Participants felt that the role of US foreign policy in the region will continue to be crucial. The perceived propping up of corrupt regimes by the United States in exchange for secure oil sources has in itself helped to promote continued stagnation. Disengagement is highly unlikely but would in itself have an incalculable effect.

Regarding the prospects for democracy in the region, participants felt that the West placed too much emphasis on the holding of elections, which, while important, is only one element of the democratization process. There was general agreement that if the United States and Europe can engage with and encourage reformers rather than confront and hector, genuine democracy would be achieved sooner.

Some Middle East experts argued that Washington has reinforced zero-sum politics in the region by focusing on top Arab rulers and not cultivating ties with emerging leaders in and outside the government.

Although the Middle East has a lot to gain economically from globalization, it was agreed that Arabs/Muslims are nervous that certain aspects of globalization, especially the pervasive influence of Western, particularly American, values and morality are a threat to traditional cultural and religious values.

## Europe and Eurasia

Participants engaged in a lively debate over whether a rift between the US and Europe is likely to occur over the next 15 years with some contending that a collapse of the US-EU partnership would occur as part of the collapse of the international system. Several participants contended that if the United States shifts its focus to Asia, the EU-US relationship could be strained to the breaking point.

- They were divided over whether China's rise would draw the United States and Europe closer or not.

- They also differed over the importance of common economic, environmental, and energy problems to the alliance.

In our Eurasia workshop, participants agreed that the United States has only limited influence on the domestic policies of the Central Asian states, although US success or failure in Iraq would have spillover effects in Central Asia. Countries in western Eurasia, they believed, will continue to seek a balance between Russia and the West. In their view, Ukraine almost certainly will continue to seek admission to NATO and the European Union while Georgia and Moldova probably will maintain their orientation in the same direction.

*"A counterterrorism strategy that approaches the problem on multiple fronts offers the greatest chance of containing—and ultimately reducing—the terrorist threat."*

On the more positive side, one of the likely features of the next 15 years is the greater availability of **high technology**, not only to those who invent it. As we try to make clear in our **Davos World** scenario, the high-tech leaders are not the only ones that can expect to make gains, but also those societies that integrate and apply the new technologies. For example, our scenario points up the beneficial effects of possible new technologies in Africa in helping to eradicate poverty. As we have noted elsewhere in this paper, global firms will play a key role in promoting more widespread prosperity and more technological innovation.

The dramatically altered **geopolitical landscape** also presents a huge challenge for the international system as well as for the United States, which has been the security guarantor of the post-World War II order. The possible contours as several trends develop—including rising powers in Asia, retrenchment in Eurasia, a roiling Middle East, and greater divisions in the transatlantic partnership—remain uncertain and variable.

- With the lessening in ties formed during the Cold War, nontraditional ad hoc alliances are likely to develop. For example, shared interest in multilateralism as a cornerstone of international relations has been viewed by some scholars as the basis

for a budding relationship between Europe and China.

As the **Pax Americana scenario** suggests, the transatlantic partnership would be a key factor in Washington's ability to remain the central pivot in international politics. The degree to which Europe is ready to shoulder more international responsibilities is unclear and depends on its ability to surmount its economic and demographic problems as well as forge a strategic vision for its role in the world. In other respects—GDP, crossroads location, stable governments, and collective military expenditures—it has the ability to increase its weight on the international stage.

*"For Washington, dealing with a rising Asia may be the most challenging of all its regional relationships."*

**Asia** is particularly important as an engine for change over the next 15 years. A key uncertainty is whether the rise of China and India will occur smoothly. A number of issues will be in play, including the future of the world trading system, advances in technology, and the shape and scope of globalization. For Washington, dealing with a rising Asia may be the most challenging of all its regional relationships. One could envisage a range of possibilities from the US enhancing its role as regional balancer between contending forces to Washington being seen as increasingly irrelevant. Both the Korea and Taiwan issues are likely to come to a head, and how they are dealt with will be important factors shaping future US-Asia ties as well as the US role in the region. Japan's position in the region is also likely to be transformed as it faces the challenge of a more independent security role.

*"A key uncertainty is whether the rise of China and India will occur smoothly."*

With the rise of the Asian giants, US **economic and technological advantages** may be vulnerable to erosion.

- While interdependencies will grow, increased Asian investment in high-tech research coupled with the rapid growth of Asian markets will increase the region's competitiveness across a wide range of economic and technical activity.

- US dependence on foreign oil supplies also makes it more vulnerable as the competition for secure access grows and the risks of supply-side disruptions increase.

In the **Middle East**, market reforms, greater democracy, and progress toward an Arab-Israeli peace would stem the shift towards more radical politics in the region and foster greater accord in the transatlantic partnership. Some of our scenarios highlight the extent to which the Middle East could remain at the center of an arc of instability extending from Africa through Central and Southeast Asia, providing fertile ground for terrorism and the proliferation of WMD.

Realization of a **Caliphate-like scenario** would pose the biggest challenge because it would reject the foundations on which the current international system has been built. Such a possibility points up the need to find ways to engage and

integrate those societies and regions that feel themselves left behind or reject elements of the globalization process. Providing economic opportunities alone may not be sufficient to enable the "have-nots" to benefit from globalization; rather, the widespread trend toward religious and cultural identification suggests that various identities apart from the nation-state will need to be accommodated in a globalized world.

The interdependence that results from globalization places increasing importance not only on **maintaining stability** in the areas of the world that drive the global economy, where about two thirds of the world's population resides, but also on helping the poor or failing states scattered across a large portion of the world's surface which have yet to modernize and connect with the larger, globalizing community. Two of our scenarios—**Pax Americana** and **Davos World**—point up the different roles that the United States is expected to play as security provider and as a financial stabilizer.

**Eurasia**, especially Central Asia and the Caucasus, probably will be an area of growing concern, with its large number of potentially failing states, radicalism in the form of Islamic extremism, and importance as a supplier or conveyor belt for energy supplies to both West and East. The trajectories of these Eurasian states will be affected by external powers such as Russia, Europe, China, India and the United States, which may be able to act as stabilizers. Russia is likely to be particularly active in trying to prevent spillover, even though it has enormous internal problems on its own plate. Farther to the West, Ukraine, Belarus, and Moldova could offset their vulnerabilities as relatively new states by closer association with Europe and the EU.

Parts of **Africa** share a similar profile with the weak states of Eurasia and will continue to form part of an extended arc of instability. As the hypothetical **Davos World** scenario suggests, globalization in terms of rising commodity prices and expanded economic growth may be a godsend where good governance is also put in place. North Africa may benefit particularly from growing ties with Europe.

**Latin America** is likely to become a more diverse set of countries: those that manage to exploit the opportunities provided by globalization will prosper, while those—such as the Andean nations currently—that do not or cannot will be left behind. Governance and leadership—often a wild card—will distinguish societies that prosper from those that remain ill-equipped to adapt. Both regions may have success stories— countries like Brazil or South Africa— which can provide a model for others to follow. The United States is uniquely positioned to facilitate Latin America growth and integration stemming the potential for fragmentation.

In that vein, the number of **interstate and internal conflicts** has been ebbing, but their lethality and potential to grow in impact once they start is a trend we have noted.

- While no single country looks within striking distance of rivaling US military power by 2020, more countries will be in a position to contest the United States in their regions. The possession of chemical, biological, and/or nuclear weapons by more

countries by 2020 would increase the potential cost of any military action by the United States and its coalition partners.

- Most US adversaries, be they states or nonstate actors, will recognize the military superiority of the United States. Rather than acquiesce to US force, they will try to circumvent or minimize US strengths and exploit perceived weaknesses, using asymmetric strategies, including terrorism and illicit acquisition of WMD, as illustrated in the **Cycle of Fear** scenario.

*"…no single country looks within striking distance of rivaling US military power by 2020."*

As our **Pax Americana** scenario dramatizes, the United States probably will continue to be called on to help manage such **conflicts** as Palestine, North Korea, Taiwan, and Kashmir to ensure they do not get out of hand if a peace settlement cannot be reached. However, the scenarios and trends we analyze in the paper suggest the challenge will be to harness the power of new players to contribute to global security, potentially relieving the United States of some of the burden. Such a shift could usher in a new phase in international politics.

- China's and, to a lesser extent, India's increasing military spending and investment plans suggest they might be more able to undertake a larger security burden.

- International and regional institutions also would need to be reformed to meet the challenges and shoulder more of the burden.

Adapting the international order may also be increasingly challenging because of the growing number of other **ethical issues** that have the potential to divide worldwide publics. These issues include the environment and climate change, cloning and stem cell research, **potential biotechnology and IT intrusions into privacy**, human rights, international law regarding conflict, and the role of multilateral institutions.

Many ethical issues, which will become more salient, cut across traditional alliances or groupings that were established to deal mainly with security issues. Such divergent interests underline the challenge for the international community, including the United States, in having to deal with multiple, competing coalitions to achieve resolution of some of these issues.

- Whatever its eventual impact or success, the Kyoto climate change treaty exemplifies how formerly nontraditional policy issues can come to the fore and form the core of budding new networks or partnerships.

- The media explosion cuts both ways: on the one hand, it makes it potentially harder to build a consensus because the media tends to magnify differences; on the other hand, the media can also facilitate discussions and consensus-building.

The United States will have to battle world public opinion, which has dramatically shifted since the end of the Cold War. Although some of the current **anti-Americanism**[13] is likely to lessen as globalization takes on more of a non-Western face, the younger generation of leaders—unlike during the post-World War II period—has no personal recollection of the United States as its "liberator." Thus, younger leaders are more likely than their predecessors to diverge with Washington's thinking on a range of issues.

Finally, as the **Pax Americana** scenario suggests, the United States may be increasingly confronted with the challenge of managing—at an acceptable cost to itself—relations with Europe, Asia, the Middle East and others, absent a single overarching threat on which to build consensus. For all the challenges ahead, the United States will nevertheless retain enormous advantages, playing a pivotal role across the broad range of issues—economic, technological, political, and military—that no other state can or will match by 2020. Even as the existing order is threatened, the United States will have many opportunities to fashion a new one.

---

[13] The Pew Research survey of attitudes around the world revealed sharply rising anti-Americanism, especially in the Muslim world, but it also found that people in Muslim countries place a high value on such democratic values as freedom of expression, freedom of the press, multiparty political systems, and equal treatment under the law. Large majorities in almost every Muslim country favor free market economic systems and believe that Western-style democracy can work in their own country.